The Soviet environment has reached crisis point – Belorussia and the Ukraine have, as a result of the Chernobyl accident, been declared ecological disaster zones, and across the country as a whole as many as 20 per cent of the population live in environmental danger areas and another 35–40 per cent in unsatisfactory conditions. According to a Supreme Soviet Environment Committee report of 1989, 80 per cent of all illness in the USSR relates either directly or indirectly to environmental problems. Not surprisingly, environmental problems have become a subject of immense public concern and have provided an anti-government, anti-party and sometimes anti-Russian catalyst.

In this timely book, leading specialists from both the West and the Soviet Union present a comprehensive and up-to-date analysis of these problems. The contributors examine the aftermath of Chernobyl, the catastrophic causes and effects of the Aral Sea's shrinkage, the environmental plight of the indigenous tundra peoples and the relationship between environmental issues and public unrest. Other chapters explore the domestic and international problems of regulation and assess the effects of perestroika and glasnost on the environment as well as on environmental politics. The depth of analysis in this volume together with the breadth of topics addressed will ensure that it is read by students and specialists of the Soviet Union and environmental issues, as well as by all government officials, journalists and industrialists with an interest in the Soviet environment.

The Soviet environment: problems, policies and politics

Selected papers from the Fourth World Congress
for Soviet and East European Studies
Harrogate, July 1990

Edited for the
INTERNATIONAL COMMITTEE
FOR SOVIET AND EAST EUROPEAN
STUDIES

General Editor
Stephen White
University of Glasgow

The Soviet environment: problems, policies and politics

Edited by
John Massey Stewart

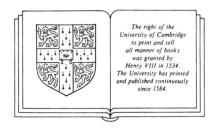

CAMBRIDGE UNIVERSITY PRESS
Cambridge   New York   Port Chester
Melbourne   Sydney

CAMBRIDGE UNIVERSITY PRESS
Cambridge, New York, Melbourne, Madrid, Cape Town, Singapore, São Paulo, Delhi

Cambridge University Press
The Edinburgh Building, Cambridge CB2 8RU, UK

Published in the United States of America by Cambridge University Press, New York

www.cambridge.org
Information on this title: www.cambridge.org/9780521117487

© Cambridge University Press 1992

This publication is in copyright. Subject to statutory exception
and to the provisions of relevant collective licensing agreements,
no reproduction of any part may take place without the written
permission of Cambridge University Press.

First published 1992
This digitally printed version 2009

A catalogue record for this publication is available from the British Library

Library of Congress Cataloguing in Publication data
The Soviet environment: problems, policies, and politics / edited by John
Massey Stewart.
    p.   ca.
"Selected papers from the Fourth World Congress for Soviet and East
European Studies, Harrogate, July 1990."– –P.
  Includes index.
  ISBN 0 521 41418 0
  1. Environmental policy–Soviet Union–Congresses. 2. Green
movement–Soviet Union–Congresses. 3. Soviet Union–Politics and
government–1985– –Congresses. 4. Pollution–Environmental aspects–
Soviet Union–Congresses. 5. Nationalism–Soviet Union Congresses.
I. Stewart, John Massey. II. World Congress for Soviet and East European
Studies (4th: 1990: Harrogate, England)
HC340.E5S68   1991
363.7'00947–dc20   91–22136 CIP

ISBN 978-0-521-41418-0 hardback
ISBN 978-0-521-11748-7 paperback

Contents

Notes on contributors	page ix
Preface	xiii
1 Environmentalism and nationalism: an unlikely twist in an unlikely direction MARSHALL I. GOLDMAN	1
2 The environmental basis for ethnic unrest in the Baltic republics PHILIP R. PRYDE	11
3 Political participation, nationalism and environmental politics in the USSR CHARLES E. ZIEGLER	24
4 BAM after the fanfare: the unbearable ecumene VICTOR L. MOTE	40
5 The massive degradation of ecosystems in the USSR ZEEV WOLFSON	57
6 The new politics in the USSR: the case of the environment JOAN DEBARDELEBEN	64
7 Water management in Soviet Central Asia: problems and prospects PHILIP P. MICKLIN	88
8 Perestroika: how it affects Soviet participation in environmental cooperation ELENA NIKITINA	115

9 US–Soviet cooperation for environmental protection: how successful are the bilateral agreements?
   KATHLEEN E. BRADEN   125

10 US–USSR nuclear safety cooperation: prospects for health and environmental collaboration
   MICHAEL CONGDON   150

11 The global impact of the Chernobyl accident five years after
   ZHORES MEDVEDEV   174

12 Glasnost, perestroika and eco-sovietology
   IGOR I. ALTSHULER, YURI N. GOLUBCHIKOV,
   RUBEN A. MNATSAKANYAN   197

13 Environmental issues in the Soviet Arctic and the fate of northern natives
   ALEXEI YU. ROGINKO   213

14 Air and water problems beyond the Urals
   JOHN MASSEY STEWART   223

Index   238

Contributors

IGOR I. ALTSHULER was a research associate of the Department of Geography, Moscow State University from 1972 to 1990, and is co-author of several books on global and regional environmental problems (atmospheric pollution, acid rains and bio-geochemical cycles). In recent years he has specialised in the USSR's environmental problems. He is co-founder of Moscow State University's Youth Council on Nature Protection (1974), the Association for the Support of Ecological Initiatives (1988) and the Independent Ecologists' Foundation (1990). Since 1991 he has been coordinating the 'Chernobyl' project of WISE (World Information Service on Energy), Amsterdam.

KATHLEEN E. BRADEN is an associate professor of geography at Seattle Pacific University. She has participated in the US–USSR Environmental Agreement under Area V for a project on snow leopard conservation techniques. She is co-author of *The Disappearing Russian Forest: Dilemma in Soviet Resource Management* (1988).

MICHAEL BRUCE CONGDON is a former US Foreign Service officer. From 1987 to 1990 he was affiliated to the US Nuclear Regulatory Commission. He is currently advisor to the Director, Division of Nuclear Safety, International Atomic Energy Agency, Vienna. From 1981 to 1984 he was distinguished visiting professor at the US Air Force Academy. In October 1990 he visited Chernobyl and the contaminated areas with the IAEA team assessing the health effects of the accident.

JOAN DEBARDELEBEN is an associate professor of political science at McGill University, specialising in Soviet domestic politics. Her publications include *The Environment and Marxism Leninism: The Soviet and East German Experience* (1985), as well as numerous articles in scholarly journals. She is editor of *To Breathe Free: The Environmen-*

tal Crisis in Eastern Europe (forthcoming) and co-author of *European Politics in Transition* (1987, new edition forthcoming 1992).

MARSHALL I. GOLDMAN is Professor of Economics, Wellesley College, and Associate Director of the Russian Research Center, Harvard University. He is the author of *The Spoils of Progress: Environmental Disruption in the Soviet Union* (1972) and *What went wrong with Perestroika* (1991).

YURI N. GOLUBCHIKOV is a research associate of the Department of Geography, Moscow State University, and a specialist in a broad range of geographical and environmental problems, especially in mountainous and Arctic territories. He is the author of several books, including *Mountains* (1988, in Russian), and is a co-founder of the Independent Ecologists' Foundation (1990).

JOHN MASSEY STEWART is a freelance writer, journalist and lecturer, specialising on Siberia and the Soviet Far East. He is at present writing a book on the natural history, geography and environmental problems of that region, *The Natural History of Russia* (1992), and is the author of *Across the Russias*.

ZHORES MEDVEDEV is a biologist who worked as senior scientist in Soviet research institutes from 1951 to 1972. Since 1973 he has been attached to the National Institute for Medical Research, Mill Hill, London. He has written more than 200 papers and articles as well as thirteen books, some with his twin brother, the historian Roy Medvedev, including *The Nuclear Disaster in the Urals* (1979), and *The Legacy of Chernobyl* (1989).

PHILIP P. MICKLIN is a professor in the Geography Department at Western Michigan University, Kalamazoo, Michigan, USA. He has studied and written extensively on water management problems in the USSR for twenty-five years, focusing primarily on large-scale water transfer proposals, water management in Central Asia, and the shrinking of the Aral Sea.

RUBEN A. MNATSAKANYAN is a research associate of the Department of Geography, Moscow State University, and a specialist on the USA's agriculture and related environmental questions and in recent years on the environmental problems of the USSR. He is a co-

Contributors xi

founder of the Association for the Support of Ecological Initiatives (1988) and the Independent Ecologists' Foundation (1990).

VICTOR L. MOTE is an associate professor of geography and Russian studies, University of Houston, USA (1971–). He is an MA and PhD (INDEA Title IV & VI Fellow), University of Washington, Seattle (1967–71), was Captain in the US Marine Corps (Vietnam Veteran) (1964–67). He is also the author of over 100 published writings, primarily dealing with Siberia.

ELENA NIKITINA is a senior researcher for the Institute of World Economy and International Relations of the USSR Academy of Sciences. She specialises in the problems of environmental security, international environmental cooperation and environmental management in the USSR. She has published a book, *World Meteorological Organisation and the World Ocean*, and is the author of about forty articles and chapters in monographs, both in Russian and English.

PHILIP R. PRYDE is an environmental analyst in the Department of Geography at San Diego State University. He specialises particularly on the USSR where he has travelled widely. He serves on the editorial board of *Soviet Geography*, has written over fifty articles and chapters in monographs, and is the author of *Non-conventional Energy Resources, Conservation in the Soviet Union* (1972), and the recent *Environmental Management in the Soviet Union* (1991).

ALEXEI YU. ROGINKO is a research associate of the Section of Environment and Ocean Development in the Institute of World Economy and International Relations (IMEMO), USSR Academy of Sciences, Moscow. A graduate of the Geographical Faculty of Moscow State University, he has published about thirty works on international marine and Arctic environmental protection issues.

ZEEV WOLFSON is editor of and frequent contributor to *Environmental Policy Review: the Soviet Union and Eastern Europe*, produced by the Centre for Soviet and East European Studies, Hebrew University of Jerusalem. He is also environmental advisor to the Israel Environmental Protection Service. He has a PhD in geography from Moscow State University and, under the pseudonym of Boris Komarov, wrote *The Destruction of Nature in the Soviet Union* which was translated into seven languages and won him Italy's 1983 Gambrinas Award for the best book on ecology.

CHARLES E. ZIEGLER is an associate professor of political science at the University of Louisville, in Kentucky. He is the author of a number of articles and monographs on the Soviet Union, including *Environmental Policy in the USSR* (1987, 1990). He is currently working on a book on Soviet relations with North-east Asia.

Preface

The Third World Congress of Soviet and East European Studies, held in Washington, DC, in 1985, contained four panels on the environment. Despite many official invitations, not a single Soviet delegate was present. Five years later, at the Fourth World Congress in Harrogate, England, the environmental panels had doubled to eight and included papers by seven Soviets: a reflection both of the increased interest by Western academia in the Soviet environment and the increased freedom under perestroika.

The fourteen chapters in this volume are a selection of the environmental papers given at Harrogate by representatives of five nationalities: Soviet, British, American, Canadian and Israeli. In some cases the papers have been altered or expanded; in others, they are new. Together, they present an illuminating and often disturbing picture of the Soviet Union's many environmental problems as well as of the policies and politics involved. To the editor's knowledge, this is the first book to give an overall picture of the Soviet environment by both Western and Soviet specialists – and, significantly, the latter do not mince their words. It is pertinent to have not only the Academy of Sciences of the USSR represented here (Nikitina and Roginko) but also one of the country's many new independent organisations (founded indeed by Altshuler and Mnatsakanyan).

Those Western authors published here include almost all the best-known names in the field of Soviet environmental studies. Among them are Marshall Goldman, who can perhaps be said to have pioneered the subject, Philip R. Pryde, whose book on Soviet nature conservation has remained the only one in the field for many years, and Philip P. Micklin, for long the West's leading expert on the (so far aborted) Siberian river-reversal scheme and now the foremost Western authority on the massive Aral Sea problem. All the authors are well-known specialists in their various spheres.

Two noted ex-Soviet citizens are also significant contributors: the biologist Zhores Medvedev who, after much painstaking research, first revealed the 1957 nuclear accident in the Urals to an (initially) highly sceptical world and has now done much work on Chernobyl, and Zeev Wolfson who, under the pseudonym Boris Komarov, sent for publication in the West a seminal work, *The Destruction of Nature in the Soviet Union*, the first full-scale account of Soviet environmental degradation.

Several environmental themes emerged clearly at Harrogate and are articulated in these pages. One was the disastrous ecological situation in the USSR and its impact not only on public health but on nationality problems. (This important ethnic dimension is discussed here by Goldman, Pryde and Ziegler.) A second theme was the effect of perestroika on the environmental situation, both for better and for worse, and the public debate and widespread concern now being voiced under glasnost. A third was the increasing Soviet participation in bilateral and multilateral environmental cooperation (see Braden, Congdon and Nikitina).

Quite apart from the major themes, many new and challenging points are advanced. Philip P. Micklin, for instance, notes two important changes in Central Asia's management of water, so crucial in this desert area: the introduction of irrigation water charges – which will hopefully end the waste of so much water – and at least a partial shift from high water-consuming crops such as cotton and rice to vegetables and other lower users. But he notes that the enormous (and notorious) water transfer project to Central Asia may be revived, pointing both to a 1988 top-level decree which directed that the scientific study of north–south water transfers continue, and to a 1990 joint declaration by the presidents of the four Central Asian republics and Kazakhstan urging the necessity of the scheme. But perhaps the strongest statement comes from Zhores Medvedev who asserts that, but for a weak bottom plate in the Chernobyl reactor which acted as an unplanned safety valve, the world's worst nuclear accident could have been an infinitely worse disaster with the meltdown of the site's other reactors and the emission of not millions but billions of curies.

This book appears at a particularly crucial time. The USSR can be said to have reached an environmental crisis. In 1990, for instance, following the example of Byelorussia, the Ukrainian Supreme Soviet declared the whole of the Ukraine an ecological disaster zone due to the Chernobyl accident. But across the whole country as many as 20 per cent of the population live in environmentally dangerous areas and another 35–40 per cent in unsatisfactory conditions and, according to a Supreme Soviet Environmental Committee report of 1989, 80 per cent of the population's diseases relate, directly or indirectly, to environmental problems. Muscovites named pollution as their principal worry in a poll taken by the city's Institute of Sociology in 1990. The impact of pollution on the air the citizens breathe, on their water supplies, on their health and their lifespan, on the very survival of the small peoples of the north (*vide* Roginko) – and, significantly, the disclosure under glasnost of the serious situation – has acted as a catalyst for perhaps millions in their attitude towards the ruling apparatus as well as serving to focus anti-Russian feeling.

Public opinion is now a political factor. According to Philip R. Pryde's paper, 331 environmental groups had come into being by 1990 and the country's first Green Party emerged in Latvia the same year. An inter-republican Green Party held an inaugural congress in Moscow in the summer of 1990 and a nation-wide green alliance is now in existence and is likely to grow in numbers and influence.

But there is a huge number of problems to be solved. Some of these problems are described in the first annual state report on the environment – a welcome new development – published in 1989 by Goskompriroda, the State Committee for Nature Protection. This committee was set up to centralise the environment's management under, moreover, the first Soviet minister not to be a party member, Professor Nikolai Vorontsov. Although the committee has now been upgraded into a ministry, it remains to be seen whether it will survive if decentralised republican bodies take over.

Gorbachev's revolution has taken us a long way from the time when pollution was regarded as a capitalist problem, when any talk of Soviet pollution was regarded as anti-Soviet and when (in 1984) the dean of the geography faculty of Moscow University, Alexander Ryabchikov, could say: 'The planned nature of our socialist economy enables us to foresee things and to take timely measures to abate the harmful effects man's economic activities have on the environment'. Gorbachev's own speech to the Communist Party Congress of August 1990 sounded the difference: 'The abandoned state of our farms, the disastrous situation with our forests and rivers, the massive ecological problems – are these not the result of the policies followed in past decades?'

I wish to thank Michael Holdsworth of CUP for his patience, Stephen White, series editor, for his valued expertise and help, Con Coroneos, copy editor, for his diligence, Tony French for his experienced advice, Jane Gowman for her indefatigable word-processing, and my wife, Penelope, for her unfailing understanding and support.

1 Environmentalism and nationalism: an unlikely twist in an unlikely direction

Marshall I. Goldman

Whatever the ultimate fate of Mikhail Gorbachev's policy of glasnost, there is no doubt that, for a brief time at least, this policy produced profound changes in the daily life of the Soviet Union. Glasnost unleashed an assault against seventy years of repression and grievances. It brought to the surface scores of accumulated slights and injustices. What was once accepted in silence suddenly was openly damned. Almost nothing remained sacred.

Of all these resurfaced issues that Gorbachev has had to deal with, the resurgence of nationalism and ethnic tension has probably been one of the most difficult to resolve. Nor does there seem to be any end in sight to the demands for secession and independence. To many of us outside the Soviet Union, such discord is particularly surprising given the relative harmony that seemed to prevail prior to 1985. If anything, Soviet propaganda often pointed to the blending of over 100 different nationality groups in the Soviet Union as a model for other societies, particularly the United States.

Although as yet not as confrontational an issue, Soviet pollution has also begun to command more and more of Gorbachev's time. However, while concern over pollution may not have provoked the sort of interethnic violence associated with nationalism, in many ways the pollution problem has served as a catalyst for the nationalist struggles. Before 1985 and Gorbachev's election as General Secretary, pollution was a relatively minor concern. Soviet officials and academics insisted that for the most part, pollution was a capitalist, not a socialist, problem. It was the inevitable result of private corporations pushing off their costs onto the public sector. Since the Soviet Union had no private corporations, by definition there could be no pollution. The public sector in the Soviet Union would not knowingly pollute itself and, after all, everything was part of the public sector. But Marxist theory notwithstanding, with time the physical evidence of the growing pollution problem became harder and harder to explain away. Alarmed over what was happening to their surroundings as well as to their health, many of those affected began to speak and write about their concerns.

The growing protest over the abuses served as a rallying point for those whose nationalist aspirations had long been dormant. In fact, almost all the nationalist movements that Gorbachev now must deal with trace their origins to early protests about the environment. The coming-of-age of environmental concerns and the newly-liberated age of glasnost combined to give rise to heretofore suppressed nationalist yearnings. Equally striking, the merger of the two interest groups occurred independently and simultaneously in several of the republics, without any evident effort at coordination.

This coincident merging of environmentalism with nationalism has given the Soviet environmental movement a character that sharply distinguishes it from the environmental movements that exist outside the Soviet Union. Inside the Soviet Union, environmentalists tend to be separatists. They are critical of Moscow and want their republics and their regions to secede from the Soviet Union. In contrast, the environmental cause in most of the rest of the world tends to be anti-nationalistic, almost 'one world' in outlook. Non-Soviet environmentalists recognise the universality of the problem and attempt to link their efforts in order to enhance their effectiveness. In seeking to find why Soviets are particularistic rather than universalistic as are their non-Soviet counterparts, we must first explain why the nationalist movement inside the Soviet Union seemed to evolve out of the environmentalist movement.

The environment in the pre-Gorbachev era

That Soviet critics seem to be so damning in their attitude towards almost anything coming out of the Soviet Union is undoubtedly a reaction to the lies that the Soviet people were told and that they themselves accepted and repeated in the pre-Gorbachev era. Whether it be the state of the Soviet economy or previous Soviet foreign policy, it almost seems that no one has anything good to say these days about the past or even present Soviet behaviour. It almost appears as if those intellectuals are attempting to atone for their past complicity with the system by bending over backwards to complain about present conditions so that no one will remember their past distortions.

Those associated in some way with either creating or eliminating pollution are among those who are especially guilt-ridden. Claims that the Soviet Union was pollution-free were cloaked in the most elegant charade. From the time of Lenin until the late 1970s, Soviet authorities prided themselves on the environmental laws enacted by the Soviet state. They were far more stringent than most comparable laws elsewhere. As one example, whereas the United States Clean Air Act of 1970 set maximum levels of carbon monoxide at 10 milligrams per cubic metre in an eight-hour period and 40 milligrams per cubic metre in a one-hour period, the maximum level set by Soviet law was only 6 milligrams per cubic metre.[1] When asked in a candid conversation if the Soviet Union enforced such laws, a Soviet official told me: 'No, enforcement of such standards would cripple all industrial production and municipal life.' But the laws were nonetheless imposed, because they were 'a sign of what a socialist system can do'.

Soviet environmental protest groups were used in the same cynical way. The main voluntary conservation group in the Soviet Union was called the Society for the Protection of Nature. As a measure of how serious the Soviet people were in their concern about conservation, Soviet authorities prided themselves on the fact that in the Russian republic alone, there were 19 million members of the society. That made it the largest environmental association in the world. What was not made explicit, however, was that many of the members did not join voluntarily, but simply had their names added to the membership roster without their permission. As one 'volunteer' put it, 'on payday you get receipts from volunteer societies and your pay is reduced accordingly ... but I'm not in that society! It's not a matter of the money. What is offensive is the deceptive and arbitrarily inflated membership totals'.[2]

There is also good reason to question just how earnest the society's leadership has been in the past. It turns out that for many years, the president of the Society for the Protection of Nature in the Russian republic was Nikolai Ovsyannikov, who in real life was also the first deputy minister of the Ministry of Land Reclamation and Water Management of the Russian republic. This was the ministry mainly responsible for the extensive irrigation and drainage programme in the Soviet Union. With the advent of glasnost, it is now revealed that this ministry 'by spending 150 billion roubles in a twenty-year period spoiled millions of hectares of the country's best lands' and was one of the most blatant disrupters of the ecological balance in the country.[3]

Against this backdrop it was not surprising that Soviet authorities did not overly concern themselves with public opinion. The contrast with Western government and industry was striking. In the West, any laws affecting the ecological balance were all but certain to draw a protest or demonstration. Until glasnost, Soviet officials had no such concerns. Unlike their Western counterparts, Soviet officials were basically free to locate Soviet factories, even nuclear-generating facilities, wherever they wanted. This helps to explain why they decided to build nuclear plants near the city centres of Gorky and Voronezh.[4] These plants were to generate steam heat and while it would be safer to build them far from the city, it would also be wasteful, and that would cost money. But since by definition Soviet scientists would never have built anything that might have been harmful, these plants were located within a few miles of the city centres and there were no protests.

Glasnost and the environment

At first, Gorbachev's coming to power did not produce much change in the handling of environmental matters. Gorbachev began to call for glasnost relatively early, but as he seemed to conceive it, glasnost was to be applied to governmental and economic affairs. The focus was intended to be on political corruption and abuse and industrial inefficiency and incompetence. The intent was to make the Soviet Union a more effective political and economic force. The last thing Gorbachev wanted was that the Soviet Union should become internally divided or a society beset by obstructionists.

That glasnost had its limits was dramatically demonstrated after the explosion at the Chernobyl nuclear plant on 26 April 1986. For three days the explosion was kept secret from the public both without and even within the Soviet Union. Gorbachev himself kept silent about what happened for a full eighteen days after the catastrophe. As we now know, four years after the fact, Soviet authorities continue to withhold vital information about the seriousness of what happened. As a consequence, as late as 1990, four million people were left to live in areas that had been seriously contaminated.

But while glasnost in environmental matters remains imperfect, there is no doubt that there is now much more openness than heretofore. If nothing else, the débacle at Chernobyl brought the credibility gap into the open. No longer could the word of governmental authorities, or for that matter senior scientists, be taken at face value. There was circumstantial evidence that some scientists felt deep guilt over their compliance in past cover-ups. Although his friends say he was bothered by family problems, the suicide of Valery Legasov in 1988, the senior Soviet scientist assigned to deal with the explosion at Chernobyl, seemed to reinforce the rumours about the self-doubt and bitterness circulating within the scientific community. There is no doubt that in many ways, Chernobyl created a major break with the past.

Most important of all, glasnost brought with it not only more openness but an end to the ban on unauthorised organisations. Prior to

Gorbachev, it was illegal to form or join any organisation not officially registered with and supervised by state authorities. That explains why the Society for the Protection of Nature was the only environmental organisation around. Almost overnight, thousands of non-governmental organisations were established. Most of them were dedicated to such harmless and non-government-oriented actions as sports. But as early as 1987, it was clear that such groups might become a force to contend with. During a summer visit to Moscow, I saw a crowd gathered in a small field off Kalinin Prospekt in the old part of Moscow. Upon enquiring, I found that they were protesting the imminent plan to chop down a 200-year-old elm tree in order to build an apartment house on the site. The neighbours were incensed that such an historic tree (they claimed that Lermontov had written his poetry under this very tree) was to be sacrificed for the construction of yet another concrete box. 'Sign the petition', they urged. To a visiting foreigner, this seemed more like a happening in Harvard Square than Moscow. But today the tree stands as an early warning signal about the influence such groups have come to have.

The growth of environmental opposition

Individuals espousing similar *ad hoc* environmental concerns had been expressing themselves for some time. But in almost every instance, because of the law, they were not formally organised or coordinated. One of the best examples is the *ad hoc* opposition to the building of two cellulose plants on the shores and territory of Lake Baikal. As early as 1960, naturalists working in the region began to warn about the possible consequences for the lake if such plants were built. It took another two years before such articles reached the national press and another three to four years before such protests reached a crescendo.

As in the case of Lake Baikal, such groups were not necessarily successful in their efforts. But similar protests over such concerns as the pollution of the Volga River, the location of nuclear energy plants in the Ukraine, Lithuania and Armenia, and the apparently unrestrained pollution of the air in most of the Soviet Union's industrial regions did accomplish one important thing – they established the credibility of those who were protesting. The protesters, even if their protests were no more than letters to local papers, were identified as critics who could be trusted to speak out. They openly complained about conditions affecting everyone's well-being. What did it matter if most such complaints did not constitute a threat to the well-being of the state? The point was that these critics had spoken out about issues affecting everyone, and were unafraid to do so. What mattered was that they were willing to challenge the calloused authorities, who for the most part were more beholden to Moscow than to their neighbours.

It was only natural, therefore, that when Gorbachev decided to relax the ban on non-state-controlled groups, those seeking leaders with integrity should turn to those who had already demonstrated their principles in their outspokenness about environmental issues. The protests in Armenia constitute one of the earliest and best case studies of how those concerned with environmental matters expanded their activities into the realm of local political concerns.

Armenians had been distressed for some time about economic developments in their republic. Lake Sevan, the largest body of water in the area, was regarded as one of Armenia's most cherished sites. But with the coming of industrialisation, it was decided to drain off some of the lake's water for a hydro-electric station. With time the lake shrank and its beauty suffered. While the electricity for the most part was used within Armenia, there was lingering resentment against the authorities in Moscow who were blamed for the whole idea.

Just as with the hydro-electric plant, there is in every one of the Soviet republics the conviction that any pollution stemming from industrial production is ultimately Moscow's fault. That may seem inconsistent with other expressions of resentment that insist that Moscow has not done enough to spread around the economic bounty of the country. That, of course, contrasts with the complaints of the Muscovites that the centre has spent too much money on the outlying republics. The Soviet Union is probably the only empire in which the centre complains as much as the periphery about being exploited by others in that empire.

But while the complaints about the centre's responsibility for pollution may seem unfairly biased, the complainants have a point. Much of the industrial activity in the various republics is the result of deliberate decisions by the central planners in Moscow. At one time, such dispersion of industrial activity throughout the country was considered praiseworthy – a sign that Moscow does not treat its republics only as raw material suppliers to the leaders in Moscow. Yet such factories were indeed used by Moscow as a way to promote the integration of such regions into the Soviet Union. Often these factories were built to a scale that was too large to be absorbed comfortably by the ecological and sometimes even the economic systems of the outlying areas. By oversizing these factories, the authorities in Moscow would then have to order the immigration of large numbers of Slavic workers from the Russian and Ukrainian republics. This not only provided more skilled workers to the region, but served to dilute the homogeneity of the various

Environmentalism and nationalism

dominant ethnic groups. Finally, the pollution control techniques supplied by Moscow were often so poor that it was usually inevitable that pollution would be the result. Once again, that served to divert grievances about local political conditions back to Moscow. In the case of Armenia, such grievances began with Lake Sevan and spread after Moscow authorities built the Medzamor nuclear energy plant on a seismic fault zone fifteen miles outside Yerevan. Another source of complaint was the Narit chemical plant, which did little to ensure control of its emissions in the heavily populated Yerevan valley.

Those who have been protesting over such planning mistakes have become well known among the general population in Yerevan. Not surprisingly, those protesting to Moscow about the deterioration in local environmental conditions quickly became the nucleus of other, more political groups that gradually decided to take advantage of the newly-relaxed law on the establishment of local non-government groups to move to protest long-standing ethnic and religious grievances. Armenians had been incensed for some time about the fact that their fellow Armenians, who constituted the majority of the population in Nagorno-Karabakh, had been shifted by Stalin from the control of Armenia to that of Azerbaidzhan. Even more upsetting was the Azeri policy of denying the residents of Nagorno-Karabakh the kind of access to the Armenian way of life that they used to have and replacing it instead with Azeri customs and language usage. Thus it seemed only natural that those protesting about environmental degradation in Armenia should formally organise into a committee for the defence of Nagorno-Karabakh once it became legal to form such groups.

The pattern was the same in most of the other republics. In March 1987, a television programme revealed a secret plan by Moscow authorities to open a phosphorite mine in Estonia. There had been long-standing complaints about the strip-mining of shale in the region. Such mines not only scarred the surface but resulted in the run-off of toxic waste. The announcement of the phosphorite mine served as a trigger. The protests about the environmental exploitation rapidly spread to encompass calls for more autonomy and eventually independence from Moscow. The region which had been politically quiescent in 1986 began to stir in 1987 with small demonstrations involving no more than 500 people, but by mid-1988, there were combined political and environmental protests involving several hundred thousand residents. The building of a cellulose plant in Latvia and the Ignalina nuclear plant in Lithuania and shared concerns about the fouling of the Baltic Sea precipitated similar reactions there.

The growth of environmentalism in Central Asia is also largely a reaction to Moscow's insistence that the region become more industrialised and at the same time that the Central Asian republics increase their output of cotton. In 1988, protests in Tashkent over Moscow's orders to build an electronics factory in a nearby mountainous recreation area sparked the formation of Birlik, a Moslem-focused nationalist movement.[5] It was the imposition of a cotton monoculture, however, that more than anything soured the local attitude toward Moscow. The cotton monoculture not only led to the suppression of a more varied crop profile, but it also resulted in the overuse of toxic pesticides and fertiliser. Equally disruptive, the thirst for cotton necessitated increasing quantities of water and the consequent diversion of rivers and canals. This explains the drying up of the Aral Sea and the resulting ecological upheaval.

It is striking to hear environmental activists from Central Asia spell out their concerns and the resulting animosity toward Moscow for what has happened. Davlat Khudonazarov, a deputy to the Congress of People's Deputies from Dushanbe, Tadzhikistan, has described in emotional terms how he grew up as a child in what he remembers as a natural paradise. While his memory may play tricks, the fact remains none the less that the glaciers of the Pamir Mountains provided him with the freshest of waters. Local authorities prided themselves on their ability to identify the origins of water much as Frenchmen can differentiate wines from different regions of the country. The water would flow on to the Aral Sea and evaporate into clouds which would snow upon the Pamir glaciers in order to repeat the cycle. With the coming of the Soviet government and the pressure to grow more cotton, those mountain streams were diverted from the Aral Sea to the cotton fields. Nature's cycle had been broken. In Khudonazarov's view, the only solution is to send home the Moscow overseers, end the cotton monoculture, and let Tadzhikistan return to its older, simpler, and separatist way.

The explosion at Chernobyl naturally enough caused everyone with a nuclear plant in their neighbourhood to take a renewed look at their own environment. We have already noted the protests in Armenia and Lithuania. The situation in Belorussia and the Ukraine became particularly urgent. Even though these were predominantly Slavic regions, the impact of nuclear fall-out provoked nationalist stirrings as well. The Ukrainians in particular began to complain about what they saw as the disproportionate number of nuclear power plants built in their republic. At the time of the Chernobyl explosion, the Ukraine had become the source of approximately 30 per cent of all the Soviet Union's nuclear capacity. Some nationalists in RUKH, the Ukrainian nationalist move-

Environmentalism and nationalism 9

ment which came into existence after Chernobyl, went so far as to see this as an effort to inflict eventual genocide on the Ukrainian people.

In another offshoot of the Chernobyl accident, Soviet authorities began to encounter resistance to nuclear weapons testing. In 1989, local authorities in Semipalatinsk in Kazakhstan, began to demand the cessation of nuclear weapons testing in their oblast. After forty years of such tests, even local Party officials had come to the conclusion that the tests had caused an unusually sharp increase in cancer rates.[6] However, plans to shift testing to the Arctic region led to the creation of anti-nuclear groups in Novaya Zemlya, a region heretofore not noted for its activism.[7]

Pollution and alienation

As bad as the environmental situation was in the Soviet Union, it is hard to see how concern for the environment would have merged so quickly with growing political disquiet were it not for glasnost and Gorbachev's decision to allow the formation of groups and organisations not subjected to formal state control. In addition, the increased flow of information made possible by glasnost was also important as was Chernobyl and the bungled attempt to cover up the full picture of what had happened. Similar ecological catastrophes have produced comparable reactions in the West. For example, the oil spill off the California coast in 1970 was one of the major stimuli to the passage of the Environmental Protection Act in the United States.

All such events only serve to alienate the non-Russian peoples from Moscow. The growing feeling that Moscow's economic policies were designed to make the outlying regions dependent on the centre by building larger-than-optimal industrial enterprises is an attitude that has begun to prevail throughout the Soviet Union. While the benefits of such efforts were seldom seen, the costs in the form of increased pollution and influx of Russian nationals were everywhere evident. As disruptive as it might be, the growing consciousness of how the local ecology had been deformed by such acts served to set off increased calls for secession. While others may question whether 'small is beautiful', there is no doubt that it is a growing conviction in the Soviet Union.

What is striking, of course, is that while the grievances of the minorities in the Soviet Union have grown louder and louder, there are now equally loud complaints coming from the Russians inside the Russian republic. Led by nationalist groups like Pamyat, they demand a halt to what they regard as the virtually free supply of valuable and non-renewable resources to the Soviet republics. The Russians have been selling petroleum at prices far below those prevailing in world markets in exchange for poorly-produced machinery and low-quality consumer supplies. In the process, the Russians have stripped themselves of their legacy for the future. No wonder the Russian empire has so few supporters.

Whereas the Green Movement in the rest of the world seeks to unify efforts to control waste and eliminate pollution, the different nationality groups in the Soviet Union have become increasingly separatist, because they associate growing pollution of their homelands with that very determination to carry out central planning and coordination. There is the widespread feeling that joining together inside the Soviet Union has made conditions worse. For these people, the solution to their problems is not unification and coordination as elsewhere in the world, but political separatism, and it has been the local environmental movement within the different republics which has sparked, and then spearheaded, that effort.

Notes

1. Marshall I. Goldman, *The Spoils of Progress: Environmental Disruption in the Soviet Union* (Cambridge, MA: MIT Press, 1972), p. 27.
2. *Izvestiya*, 24 February 1986, p. 6.
3. *Foreign Broadcast Information Service*–SOV 89–134S, 14 July 1989, p. 13.
4. Marshall I. Goldman, *The Enigma of Soviet Petroleum: Half Full or Half Empty?* (Winchester, MA: Allen & Unwin, 1980), p. 152.
5. *The Wall Street Journal*, 21 July 1989, p. A12.
6. Radio Liberty/Radio Free Europe, 23 July 1989, p. 34.
7. *SOVSET*, 17 April 1990, p. 4.

2 The environmental basis for ethnic unrest in the Baltic republics

Philip R. Pryde

The ascent of Mikhail Gorbachev to the leadership of the Soviet Union in 1985 engendered a movement for more candour, that is, glasnost, in the discussion of Soviet problems. Among the problems currently being more fully debated are those dealing with environmental deterioration.

This has been particularly true in the Baltic republics, where glasnost has allowed environmental concerns to be frequently and vociferously aired. These concerns are perceived by residents of the region as so significant that they are often cited as a major justification for their demands for independence.

The environmental factor in the Baltic independence movement is not as well publicised in the Western world as are the other causes of this movement. This essay will summarise the nature of these ecological concerns, and discuss some of the options available to the Baltic republics for resolving them.

Conservation efforts prior to the glasnost era

Prior to their forced annexation to the USSR in 1940, the Baltic republics had achieved an admirable conservation record. Nature reserves had been created in Estonia and Latvia, a nature conservation society has existed at Tartu University since 1920, and the three republics had passed numerous conservation laws. Even under Soviet rule, Estonia was the first Union republic to enact a comprehensive nature protection law (1957), and the first to form an independent green group (1987).

In the period between 1945 and 1985, conservation efforts – or their relative absence – in the Baltic region were fairly similar to those elsewhere in the USSR. Natural resources were harnessed for the national good, pollution was tolerated as an economic necessity, and schools taught that there could be no unwise use of natural resources under socialism.[1] Conservation efforts in the early post-war period

(1946–65) generally promoted only the more economical use of natural resources.

The forests in the region had been heavily depleted during World War II. After the war, most of the forests in these republics were placed into Group I (protected) or Group II (restricted cutting) status, rather than given commercial timber status (Group III). Sustained yield timber harvesting, which equates the cut to the new growth, is practised in the Baltic republics, although some overcutting of coniferous stands in accessible zones still occurs.[2]

Many areas in the Baltic republics have been set aside for biotic preservation purposes. The most numerous and important of the various types of preserved areas in the Soviet Union are the state nature reserves, or *zapovedniki*. These areas are set aside for both nature preservation and scientific research purposes. There are thirteen *zapovedniki* in the Baltic region: five in Estonia, five in Latvia, and three in Lithuania. They range in size from over 48,000 hectares to under 600 (table 2.1).

Zakazniki, or partially or temporarily protected areas, are another common category of preserved land in the USSR. The Lithuanian republic has 174 of these preserves, Latvia has 148, and Estonia 57.[3] The Russian republic and some others use this category primarily for game management purposes, whereas the Baltic republics have established them mainly to preserve specific features of the landscape.

In 1971, the first national parks appeared in the Soviet Union, starting with 'Lahemaa' in Estonia. These parks became models for those that followed elsewhere in the USSR. The national parks are seen as complementing the network of *zapovedniki*, which are not intended to serve a tourist function.

The first such park, Estonia's Lahemaa, takes in 64,911 hectares on the Gulf of Finland.[4] Its management zones consist of reserve (restricted) areas (6–8 per cent of the total area); landscape preservation districts where passive recreation is permitted (about two-thirds of the park); and various 'cultivated landscapes' (30 per cent). Intensive recreation zones exist around the two towns in the park. The park also contains many archaeological sites.

The second park to be created was Gauya National Park in Latvia. Located 50 kilometres from Riga along the Gauya River, it is the largest of the Baltic parks at 83,750 hectares. The park is divided into conservation zones, recreation zones, and 'neutral' zones, the latter including farms and villages. The park contains castles dating back to the thirteenth century and a great many archaeological and cultural sites.

Table 2.1. *Nature reserves in the Baltic republics*

Reserves	Area (ha.)	Year created
Estonia		
Endla	8,162	1985
Matsalu	48,634	1957
Nigula	2,771	1957
Vil'sandi	10,689	1910
Viydumyae	593	1958
Latvia		
Grini	1,076	1957
Krustkalny	2,826	1977
Moritssala	818	1912
Slitere	14,882	1921
Teychi	18,966	1982
Lithuania		
Chapkyalyay	8,469	1975
Kamanos	3,660	1979
Zhuvintas	5,428	1946

Source: G. R. F. Drucker and Z. J. Karpowicz. *Directory of Protected Areas: Eastern Europe & USSR* (Cambridge, IUCN Protected Areas Data Unit, 1989).

Recreation is most intensive around the towns of Tsesis and Sigulda, the latter hosting up to 900,000 tourists a year.[5]

The Lithuanian National Park is 30,000 hectares in size, not including its surrounding buffer zone. It is located in the republic's lake district, immediately west of the city of Ignalina. The park's sixty-two lakes take in 16 per cent of its area.[6] Its three primary management zones are scientific-conservation, recreation, and agriculture/forestry. A short distance away is the Ignalina atomic power station, one of the largest nuclear generating complexes in the USSR.

A possible (but unconfirmed) new national park may exist in the Baltic region, though not in the Baltic republics. References have been seen to a Kurshskaya Kosa National Park, situated on the Kurshsk spit which extends into the Baltic Sea near Kaliningrad, though a 1990 list of Soviet national parks supplied by Goskompriroda, the State Committee for Environmental Protection, does not include it.

One function of these preserved areas is to help protect endangered species. The USSR 'Red Book' which records such species, lists four endangered forms of wildlife in the Baltic republics: the Baltic grey seal, the Baltic race of the harbour seal, the bottlenosed whale, and the European pearl oyster.[7]

Recent environmental concerns

During the 1980s, ecological concerns in the Baltic region tended to focus most sharply around various forms of environmental pollution. Certainly a key factor here, as in much of the Soviet Union, was the 1986 Chernobyl accident. All of the Baltic region, but especially the southern half, lay in the downwind direction from the burning reactor.

Chernobyl was of particular concern in Lithuania, which contains the large Ignalina nuclear power complex. Its units, like those at Chernobyl, are of the graphite-moderated RBMK type, but the Ignalina reactors are even larger (1,500 megawatts rather than 1,000), and at present are the largest in the world. In the wake of Chernobyl, there has been considerable public opposition to the Ignalina operation, in part a response to a small fire in unit 2 at the plant in 1988. A bumper sticker has been prepared that shows a pile of skulls on one side and on the other the wording (in Lithuanian) 'Yesterday Chernobyl, tomorrow Ignalina?'. Another cause of the Ignalina protests is the voluminous discharge of heated water from the reactor's cooling cycle into nearby lakes. Irreversible ecological damage to the lakes is feared possible.

Offshore oil drilling in the Soviet Union occurs mainly in the Caspian Sea, where output is declining. However, exploratory drilling has taken place recently in the Baltic Sea.[8] Anywhere there is oil, accidents occasionally occur. In 1977, a Soviet tanker went aground in Sweden, producing the worst oil spill up to that time in the Baltic Sea. Four years later, a British tanker broke apart in a storm just offshore from the Lithuanian port of Klaipeda. The boiler fuel slick from that accident was reported to have covered about 40 kilometres of sandy beaches near resort towns to the north. It also penetrated 20 kilometres into the shallow Kurland lagoon south of Klaipeda, causing an estimated 600 million roubles worth of damage.[9]

The Baltic coastline is recognised as a major problem area. As long ago as 1976 a special resolution 'On measures for stronger protection of the Baltic Sea basin against pollution' was passed by the USSR Council of Ministers. It called for improvements in the use and disposal of municipal sewage, oil products, pesticides, and other potential pollutants. But Latvian beaches continue to be soiled. In 1988 the beach at Jurmala (near Riga) was closed due to severe bacteriological pollution; private cars could enter Jurmala by special pass only.[10] In Estonia, the

Table 2.2. *Major sources of pollution into the Baltic Sea in resort areas**

	Million m³ of polluted water per year		
Site	1985	1987	1988
City of Jurmala (Latvia)	2.4	2.9	3.2
Sloka cellulose-paper factory (Latvia)†	15.0	14.7	14.8
City of Palanga (Lithuania)	3.0	4.1	4.8
City of Parnu (Estonia)	5.1	5.7	5.5

* Figures for the city of Riga are not available.
† The Sloka mill was closed down in 1989.
Source: *Narodnoye khozyaistvo SSSR v 1988g.*, p. 248.

beach at Parnu was also closed in 1988 by wastes from nearby slaughterhouses. Cellulose factories and municipal wastes from such cities as Leningrad, Riga, and Ventspils have been cited as some of the major sources of Baltic pollution.[11] The seriousness of the Baltic Sea situation has engendered a special page of data in the 1988 USSR statistical handbook (table 2.2).

In response, the Sloka pulp and paper mill at Jurmala was ordered closed as of January, 1990. The economic consequences of this were widespread, as workers in support industries had to be laid off and Latvian newspapers had to cease publication.[12]

Polychlorinated biphenols, or PCBs, are an industrial chemical whose breakdown products strongly resemble those of DDT. They were identified as a problem in the Soviet Union in 1974 (by Swedish scientists) who discovered them flowing from the Neman River into the Baltic Sea. They were subsequently shown to be present in virtually every water body in the country, with the military identified as their main source.[13]

Air pollution is serious in all three republics, but especially so in Estonia, where it is the highest in the USSR (table 2.3). Accidents also occur: in Jonava, Lithuania, a large ammonia tank exploded in 1989, killing at least four persons and injuring dozens. The ammonia gas cloud necessitated the evacuation of 30,000 people.[14]

The consequences of possible long-term atmospheric warming should also be noted. If significant warming were to occur, it could result in (among other things) a significant melting of polar ice caps. This could ultimately raise the level of the world ocean by as much as a metre or more, flooding at very high tides the low-lying portions of countries and port cities around the world. Among the cities that could be affected in the Soviet Union are Leningrad, Riga, Tallinn, Klaipeda, and many

Table 2.3. *Comparative air pollution indices by Union republic*

Republic	Pollution in tons per capita per year*	per km² per year*	Cost of pollution control, roubles per capita
Estonia	0.67	41.0	98.0
Kazakhstan	0.59	4.7†	64.0
RSFSR	0.41	6.0†	90.0
Ukraine	0.36	31.1	66.0
Turkmenia	0.31	4.8†	8.2
Belorussia	0.21	10.6	48.0
Moldavia	0.20	24.0	24.0
Lithuania	0.19	10.6	35.0
Azerbaidzhan	0.17	12.6	26.0
Georgia	0.14	14.2†	15.0
Uzbekistan	0.12	4.8	19.0
Latvia	0.12	4.7	33.0
Armenia	0.10	23.9	3.6
Kirgizia	0.08	1.8	10.4
Tadzhikistan	0.05	3.0†	9.5
USSR ave.:	0.36	6.9†	70.0

* Pollution includes the waste products from industry and from transport. The calculations are based on *Narodnoe khoziaistvo SSSR*, 1984 Moscow, 1985, pp. 403–6; T. Khachaturov, *Ekonomika prirodopol'zovaniia*, Moscow, 1982, p. 91. It is necessary to note that the Soviet statistics do not take into account small industrial sources, the heating of homes and part of the pollution from transport. Therefore the real emissions may be higher by 20–30%.
† In these republics the calculation is not made for the whole territory but for the areas being 'economically developed' as suggested by the figures of Khachaturov.
Source: Adapted from 'Local problems . . .', *Environmental Policy Review*, June 1987, p. 25 (reprinted by permission).

smaller towns along the Baltic coast. A raised ocean level could also produce increased coastal erosion.

The mining problem in Estonia

Estonia is one of the most highly polluted regions of the Soviet Union. The major cause of this is phosphorite and oil shale mining activities.[15] Both the phosphorite, used in making fertilisers, and the oil shale, used mainly as a power plant fuel, are surface mining operations that scar thousands of hectares of land. Estonians feel their resources are being exploited for the benefit of others, since the fertilisers are of limited use

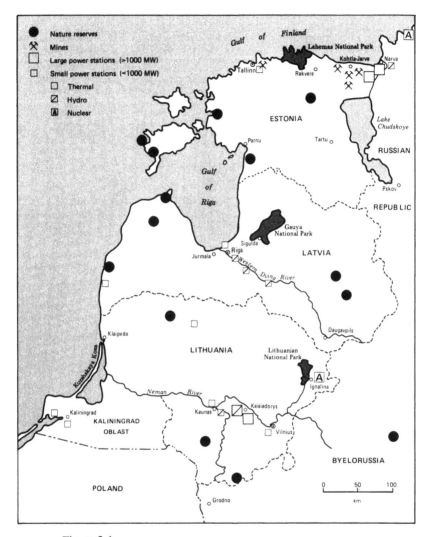

Figure 2.1

on Estonia's poor soils and much of the electricity produced from the oil shale goes outside the republic.

The mines are largely located in the heavily travelled corridor between Tallinn and Narva (figure 2.1). The surface mines are required by law to be restored to productive uses,[16] but compliance is apparently lax. Across the USSR there are numerous 'moonscapes' that seem to persist in surface mining regions, and in Estonia the mining operations have been sharply criticised by local officials. Indeed, Estonia has the greatest *per capita* amount of land destroyed by surface mining of any of the fifteen Union republics.[17]

Perhaps of even greater concern than the biosphere degradation caused by the mining operations is the pollution produced from processing the phosphorite ore, and from processing and burning the shale. One survey found that 90 per cent of the respondents in the mining centre of Kohtla-Jarve concurred that 'environmental problems are so critical and significant that immediate remedies must be taken'.[18] A cause of this reaction may be that their children are suffering. According to the newspaper *Sovetskaya Estoniya*:

> A republic commission appointed to determine why large numbers of children in north-eastern Estonia have fallen ill has come to the conclusion that the reason for the children's hair loss and other health abnormalities is, in all likelihood, the pollution of the environment and of daily living conditions in that region . . . In Sillamae, Narva, and Kohtla-Jarve, hair loss has been noted in more than 200 children. We believe that north-eastern Estonia should be considered an environmentally critical region, where no more polluting enterprises can be built . . .[19]

Another complaint reflects how the mining industry is organised in the Soviet Union. The ministries controlling mining are All-Union ministries that are run from Moscow, with industry profits going to the centralised ministry. But new industry in Estonia usually requires imported workers, mainly Russians; this is a sensitive point in a republic with a low birth rate that now consists of only 61 per cent Estonians. Also, the Estonian budget must pay for these newcomers' housing, schools, and health facilities. To this are then added the problems of land, air, and water pollution.

It is thus easy to see why the Estonians strongly resist current proposals to expand the mining operations. Indeed, they would prefer that the mining either be terminated or greatly reduced in scope. Estonian leaders have made it clear that if decisions concerning the republic's economy were made in Tallinn rather than Moscow, the mining would be quickly scaled back.

Glasnost and public environmental activism

With the advent of the glasnost era in 1986, public hesitancy to voice opinions on environmental issues quickly dissipated. Instead of merely letters to newspaper editors, the preferred mode for driving home a point became street demonstrations. As Fyodor Morgun, then chairman of Goskompriroda, put it, 'please believe me, the people have awakened'.[20]

In addition to demonstrations opposing Estonian mining and the Ignalina atomic power plant, Latvians have protested the proposed hydro-electric station at Daugavpils. Thirty thousand signatures in protest at the project were gathered, and in 1987 it was cancelled.[21] A surprising number of these demonstrations have been successful in at least forcing further review of the project.

An unusual protest took place in Estonia in 1989. The newspaper *Sovetskaya Estoniya* on June 3 reported that many Estonian men who had assisted in the Chernobyl clean-up were 'getting sick, dying'. Five were reported to have died, and at least 227 others required medical treatment. An 'Estonian Chernobyl Committee' has been formed; similar committees exist in other republics.

Since the start of glasnost, local 'green' groups have sprung up in various locations in the USSR. Each of the three Baltic republics has indigenous green groups that are quite active. In 1990, the Social-Ecological Union of the USSR, trying to function perhaps as the first nationwide conservation organisation, published a list of 331 USSR environmental groups, of which five were in Latvia, five in Lithuania, and four in Estonia. Although these organisations are relatively small at present, it is quite probable their numbers and influence will increase with time. As evidence of this, the Latvian 'greens' organised themselves in 1990 as the Green Party, the first green political party to emerge in the Soviet Union.

In 1988 the Estonian Green Movement (EGM) was created.[22] The EGM apparently does not maintain a centralised membership roster, but activists in the movement with whom I spoke in Tallinn estimated the active membership in 1989 at 4,000–5,000, with perhaps a like number of less active supporters. Although not defining itself as a political movement (the Popular Front of Estonia plays this role), the EGM works closely with the republic's political establishment, and virtually all Estonian leaders share its goals. For example, both groups feel that their republic's environmental situation would be improved if economic independence from Moscow's ministries could be attained.

A little to the north, but still in the Baltic realm, Leningraders have formed several local environmental organisations. One of their main activities has been protesting about the adverse changes to the Gulf of Finland caused by the new Leningrad flood control dike.

In recent years, these green groups appear to have explicit governmental approval of their activities. Goskompriroda is specifically authorised to assist the efforts of local environmental groups.[23] Indeed, the 1988 State of the Soviet Environment report, published by Goskompriroda, describes the non-governmental ecology organisations in the

USSR, and lists by name such groups in fourteen Soviet cities.[24] The 1990 draft Law on Nature Conservation specifically grants citizens the right to form ecology organisations and to have access to pollution and other environmental information.[25]

These organisations have even forged ties with similar green groups abroad, a highly unlikely action prior to the Gorbachev era. For example, the EGM in 1988 established a Soviet precedent by becoming an associate member of the Western environmental group, Friends of the Earth.

What has been done?

The long-standing practice of creating *zapovedniki*, national parks, and *zakazniki* as a means of promoting biotic preservation has already been noted, as has the more recent formation of local green groups. Protests by such environmental activists appear to be often effective. For example, the combination of the 1988 fire at Ignalina and the numerous public protests may have led to the cancellation of the planned third unit at that complex.

The Estonians' anti-mining protests have apparently met with some success; development of a new phosphorite mine near Rakvere has been halted, with some possibility that this activity might be phased out entirely in the 1990s.[26] This will not end the problem, however, for as noted earlier Estonia has the most land disturbed by mining *per capita* of any republic, and reclamation efforts are slow.

There has been an expanded emphasis on environmental education in recent years. Such efforts include school programmes, posters and displays in parks, exhibits in museums, and popular books on environmental topics. As one example, the Latvian Museum of Nature in Riga contains a large room of exhibits on environmental issues; it appeared to be very well done.

Less polluting forms of energy production are being sought. One of these is pumped-storage hydro-electric facilities that make use of otherwise wasted night-time generation capacity at nuclear plants to pump water uphill for later use. Among the Soviet Union's pumped storage plants is the one at Kaisiadorys in Lithuania, which has a capacity of 1,600 megawatts.

The 1976 USSR resolution on protecting the Baltic Sea basin from pollution has been noted, together with the fact that it was unable to prevent many beaches along the Baltic coast from having to be closed during the summer of 1988. Additionally, the Soviet Union is also a party to an important regional treaty involving the Baltic Sea. The necessity for international cooperation is caused by the Baltic Sea's resemblance to a large saline lake; it is open to the world ocean only via narrow passages through the Danish islands. Water circulation within it is poor, and pollution, originating in both Eastern and Western nations, pours into it continually. In 1974, all the surrounding nations signed the International Convention on the Protection of the Marine Environment of the Baltic Sea, whose secretariat is housed in Helsinki. To help implement this accord, in 1988 the signatories agreed to cut in half their discharge to the sea of nutrients, organic toxins and metallic wastes by 1995.

Options for the future

The existing political and economic institutions of the Soviet Union have never been adept at internalising the environmental costs of industrial expansion.[27] If the Baltic republics remain in some form of dependent relationship with the USSR, their environmental problems will be difficult to fully overcome. But even if they achieve political independence, the road is unlikely to be much smoother.

If full political independence should be unattainable in the foreseeable future, what the Baltic republics would like in the interim is a high degree of *economic* independence from the centralised ministries in Moscow, with the authority over all industrial decisions (and thus over major sources of pollution) residing in Tallinn, Riga, and Vilnius. Reflecting this, resolutions of 'economic independence' have been passed in all three republics. In response, the USSR Supreme Soviet in late 1989 endorsed a proposal in principle for economic independence for the three Baltic republics, but this may be in abeyance in light of the severe political repression which took place in 1990–1. Also, although economic independence is favoured in the Baltic region, the idea has been opposed by the central planning bodies in Moscow (who saw their job becoming more difficult), and by other minority republics who questioned (or perhaps wished to be part of) the preferential treatment.[28]

What the Soviet and Baltic leaders might find acceptable is to utilise the Baltic republics as centres of 'new economic thinking' and experimentation, allowing them (as part of perestroika) to be test centres for private enterprise, decentralised planning, and market economy experiments. This could be a way to implement certain goals of perestroika, while at the same time giving the Baltic states a measure of the special economic status that they desire. In an environment where the status quo is not acceptable to anyone, it might be worth a try as an interim measure.

Even if the Baltic states were to become independent, they would face difficulties in resolving their environmental problems. If Estonia closes its mines, public health would be improved, but unemployment and a significant source of foreign exchange would be lost. Non-polluting industries would quickly have to be developed to fill both functions. Although Lithuania might like to shut down the Ignalina nuclear station, this seems as unlikely to happen under independence as at present. This plant is the major power producer for all three Baltic republics (especially if oil shale mining is terminated in Estonia), and no readily available substitute for it is at hand. Its closure would also greatly reduce the economic effectiveness of the Kaisiadorys pumped-storage plant. Lithuania could also use the income Ignalina could provide from sales of electricity outside the Baltic area (in this regard, we will ignore the very tricky question of whether the USSR would require Lithuania to somehow pay for taking over the Ignalina plant).

In summary, the environmental problems of the Baltic republics, while a major driving force in their quest for independence, will by no means be quickly or easily resolved even if this dream is realised. These problems took years to develop into their present critical state, and under any political scenario they will take many years to resolve.

Notes

1. P. R. Pryde, *Conservation in the Soviet Union* (Cambridge: Cambridge University Press, 1972), p. 7.
2. B. Barr and K. Braden, *The Disappearing Russian Forest* (Totowa, NJ: Rowman and Littlefield, 1988).
3. A. Shalybkov and K. Storchevoi, 'Nature preserves: a reference guide', *Soviet Geography*, 29 (1988), pp. 589–98.
4. A. Kaasik and E. Kask, *Lahemaa zahvuspark* (Tallinn: Eesti Raamat, 1983).
5. I. I. Bergholtsas, 'What should a national park be like?', *Sovetskoye gosudarstvo i pravo*, 1976, no. 1, pp. 72–4, as translated in *Current Digest of the Soviet Press*, 28, no. 18 (1976), p. 8.
6. *Eesti loodus*, May 1975.
7. P. R. Pryde, 'The distribution of endangered fauna in the USSR', *Biological Conservation*, 42 (1987), pp. 19–37.
8. W. Kelly, H. Shaffer and J. Thompson, *Energy Research and Development in the USSR* (Durham: Duke University Press, 1986).
9. T. Shabad, 'Vast damage cited in Baltic oil spill', *New York Times*, 31 January 1982, p. 10.
10. *Izvestiya*, 17 July 1988.
11. *Doklad: Sostoyaniye prirodnoy sredy v SSSR v 1988 godu* (Report on the

state of the Soviet environment for 1988) (Moscow: Gosudarstvennyi Komitet SSSR po Okhrane Prirody (Goskompriroda), 1989), p. 59.
12. I. Litvinova, 'Why the newspapers didn't come out in Latvia', *Izvestiya*, 23 January 1990, p. 3, as translated in *CDSP*, 42, no. 4 (1990), pp. 10–11.
13. B. Komarov, *The Destruction of Nature in the Soviet Union* (White Plains, NY: M. E. Sharpe, 1980), p. 33.
14. *Izvestiya*, 21 March 1989.
15. B. Jancar, *Environmental Management in the Soviet Union and Yugoslavia* (Durham: Duke University Press, 1987), p. 173f; M. Taagepera, 'The ecological and political problems of phosphorite mining in Estonia', *Journal of Baltic Studies*, 20 (1989), pp. 165–74.
16. 'Principles of USSR and Union-republic legislation on land', *Pravda*, 7 March 1990, pp. 2–5.
17. A. Bond and K. Piepenburg, 'Land reclamation after surface mining in the USSR: economic, political, and legal issues', *Soviet Geography*, 31, no. 5 (May 1990), pp. 332–65.
18. 'Local problems (Central Asia without Siberian water; the Estonian phenomenon; Armenians protest)', *Environmental Policy Review*, 1 (1987), p. 27.
19. J. Uibu, 'Appeal of the Estonian Republic Ministry of Public Health', *Sovetskaya Estoniya*, 16 May 1989, p. 3, as translated in *CDSP*, 41, no. 20 (1989), pp. 27–8.
20. *Time*, 2 January 1989.
21. N. R. Muiznieks, 'The Daugavpils hydro station and *glasnost* in Latvia', *Journal of Baltic Studies*, 18 (Spring 1987), pp. 63–70.
22. D. Devyatkin, 'An interview with a leader of the Green Movement', *Environment*, 30, no. 10 (December 1988), pp. 13–15. *Some Facts from the First Year of EGM* (Tartu: Estonian Green Movement, 1989).
23. W. E. Freeman, 'The politics of environmental protection in the USSR: the case of the Soviet EPA', unpublished USIA Research Memorandum, dated 22 September 1989.
24. *Doklad*, ibid. pp. 207–8.
25. 'Zakon SSSR ob okhrane prirody (proekt)', (Law of the USSR on conservation of nature), draft version being reviewed by various Soviet legislative bodies, Spring 1990.
26. *Soviet Geography*, 29, no. 10 (1988), pp. 951–3.
27. J. DeBardeleben, *The Environment and Marxism-Leninism* (Boulder, CO: Westview Press, 1985); Jancar, *Environmental Management*; C. E. Ziegler, *Environmental Policy in the USSR* (Amherst, MA: University of Massachusetts Press, 1987).
28. *New York Times*, 28 July 1989.

3 Political participation, nationalism and environmental politics in the USSR

Charles E. Ziegler

Political power in the once highly-centralised Soviet Union has devolved so rapidly that former congressman Tip O'Neill's declaration about the United States – 'all politics is local politics' – is now as true of Tallinn and Perm as it is of Tampa and Portland.[1] Although the roots of a civil society emerged with the progress towards universal education and growth in mass communications, Soviet political development lagged far behind changes in the political culture. Political participation throughout most of Soviet history was highly circumscribed and, most scholars agreed, significantly different from participation in democratic systems.[2] By the late 1980s, however, popular demands were being articulated through competitive elections, massive demonstrations, petitions and the activities of thousands of independent social and political organisations.

Several factors account for the rapid emergence of qualitatively new forms of political participation in the USSR. First, the disastrous environmental record compiled by Soviet authorities, with its pervasive and highly visible impact on public health, has created the objective conditions for popular discontent. Secondly, glasnost has made possible the relatively free exchange of information and ideas necessary for effective discussion of political issues. Thirdly, democratisation has introduced the concept of official accountability and responsiveness into Soviet political life. Fourthly, the acceptance of societal and political pluralism has legitimised the aggregation of diverse interests by organisations not directly under the control of the Communist Party. Clearly, the underpinnings of participatory democracy are emerging in the Soviet Union.

This chapter explores currents of Soviet grassroots participation in the Gorbachev period by highlighting examples of ecological activism in the Russian and non-Russian republics. Environmental participation in the Soviet Union is found to be highly negative in orientation, is closely interwoven with nationalistic movements, and contains a strong populist element. Some general observations on Soviet political participation as it relates to environmentalism are advanced at the end of the chapter.

Nationalism and environmental pollution

Economic development and ecology issues are closely linked to nationalist and separatist demands in the USSR. Many ethnic minorities are convinced that environmental problems affecting their republics have resulted from imprudent development policies pursued by Moscow. Central authorities have been criticised for exploiting the republics' natural resources without regard for the ecological consequences of such development strategies. Declarations of sovereignty by the union republics and smaller national units have denied Soviet authorities legal jurisdiction over their natural resources and have in many cases demanded compensation for environmental harm inflicted on the locale. Regional control over natural resources has become a contentious issue, frustrating efforts to develop a new Union Treaty.

In the Soviet Union, nationalism and environmental issues encompass territorial, biological and psychological aspects. The legacy of central planning and Russian domination has infused environmental issues within the union republics, autonomous republics, autonomous regions and national areas with colonial, exploitative connotations. In certain regions, the ecological situation has become so serious that minorities believe the damage to health threatens the biological viability of their national group. Economic development in most republics has been accompanied by an influx of Russian workers, who not only command many of the better urban jobs and dilute the political strength of ethnic minorities but also symbolise Moscow's penetration of the republics. Environmental pollution has become a conspicuous example of Russian domination and bureaucratic unresponsiveness to minority concerns. Since the list of environmental tragedies in the minority republics is far too long to examine in detail, the following discussion will be confined to the more noteworthy instances.

The non-Russian nationalities and environmentalism

In Central Asia, the best-known and most tragic environmental disaster is the disappearing Aral Sea, whose feeder rivers, the Amu-Dar'ya and Syr-Dar'ya, were depleted as water was diverted to irrigate cotton fields.[3] In thirty years, from 1960 to 1990, the Aral Sea lost more than 40 per cent of its surface area. As the shoreline retreated, up to 75 million tons of wind-blown salts annually have been deposited onto the surrounding communities, causing marked increases in throat cancer, respiratory and eye diseases, dysentery, and hepatitis. Drinking water has been contaminated over a vast area around the sea, and infant mortality is the highest in the USSR. According to one source, 90 per cent of children born in the Karakalpak autonomous republic suffer from anaemia, as do 25.4 per cent of children in the Uzbek republic.[4]

Responsibility for the disastrous environmental record in the Aral basin rests primarily with Moscow. Under Brezhnev, the central government pushed local authorities to expand cotton production, the export of which provided an important source of convertible currency. Corrective measures taken in the post-Brezhnev period have not proved effective. Although Party and government resolutions have mandated an end to the use of DDT and the defoliant Butifos on cotton fields, use of these highly toxic chemicals continues to appear in mothers' milk. According to a spokeswoman for the informal Uzbek political group Birlik, an average 54 kilograms of chemicals are used on each hectare of Uzbek land every year.[5]

In general, informal groups have not played as prominent a role in Central Asian politics as they have in other regions of the Soviet Union. Birlik, established as a small working group by Uzbek intellectuals in November 1988, is among the better-organised Central Asian initiative groups. This group has called for an end to cotton monoculture in Uzbekistan and has pressed to have Uzbek designated the official state language of the republic. Uzbek authorities have for the most part approved of environmental groups' efforts to save the Aral Sea, but have criticised Birlik for reputedly exacerbating inter-ethnic tensions.[6]

Responding to public pressures to diversify Central Asian agriculture, the central Communist Party Politburo in September 1988 adopted guidelines directing regional authorities to reduce the area devoted to cotton to a maximum of 70 per cent and to take measures to conserve water. Given the country's precarious economic situation, Moscow cannot provide funds required to remedy the situation; the capital investment needed to diversify agriculture in the affected republics of Uzbekistan, Tadzhikistan and Kazakhstan is simply not available.[7] Regional officials continue to urge that work on the Siberian water diversion projects be resumed as a solution to the shrinking sea, a position opposed by local ecologists and Russian nationalists.

In the Caucasus, the capital cities of both Armenia and Azerbaidzhan are heavily polluted. Air quality in Yerevan has been affected by the expansion of motor vehicle transportation and the continued growth in the number of chemical plants within city limits, resulting in higher levels of birth defects and infant mortality rates and increased sterility rates among adults. Public discontent over the pollution of Lake Sevan (largely from agricultural run-off), the Armenian nuclear power plant at Metsamor (closed after the Spitak earthquake), pollution of the Razdan River (compounded by hydro-electric development projects), led to mass ecology demonstrations in late 1987.[8]

Air pollution in Baku, the capital city of Azerbaidzhan, is reportedly among the worst in the Soviet Union, and has significantly raised morbidity levels in the republic.[9] The Caspian Sea is heavily polluted by untreated sewage from Baku and from oil and sediment discharged into the sea by petroleum industries in the capital and Sumgait. Agriculture in the republic is subject to over-chemicalisation: according to the Azerbaidzhan Minister of Health, an average 40 kilograms of pesticides is used on each hectare of cotton and vegetable land and 183 kilos in the grape-producing regions. Other researchers have found anaemia rates in children below the age of fourteen two and a half times the average rate for the USSR. Of the region's women between the ages of twenty and thirty-four, 42 per cent are reportedly sterile.[10]

Nationalistic forces in the republic have focused many of their demands around these environmental issues. In early 1989 the Azerbaidzhan Popular Front disseminated a draft programme stressing the importance of addressing the ecological crisis and demanding that the republic be given full control over its natural resources. The programme condemned Moscow for its 'barbaric exploitation' of Azerbaidzhan's natural resources and blamed the central leadership for the ecological disaster facing the republic.[11]

Moldavia, like the Central Asian republics, was pressured by the Brezhnev leadership into debasing its rich agricultural soil through over-chemicalisation. While the Soviet average for application of pesticides is reportedly 0.5 to 1.0 kilograms per hectare, roughly comparable to developed Western countries, the average in Moldavia and Central Asia is 20.6 kilograms or higher.[12] The land has become impoverished, wildlife has been decimated, and drinking water is in short supply. Carelessly stored chemicals have contaminated lakes and rivers. Every year tens of thousands of agricultural workers handling pesticides fall ill, and Moldavian children have been found to suffer from lowered intelligence levels.[13]

A Moldavian Green movement, initially opposed by the Communist Party organisation, was established in November 1988 by a group of Kishinev intellectuals. The Moldavian Greens have directed their efforts towards addressing environmental problems in agriculture, nuclear power, and pollution of the Prut river. They have been assisted by other Moldavian groups, including a Public Committee to Save the

Prut River, based in Kishinev. This organisation seeks to enlist the support of Roumanian scientists in attempting to clean up the heavily polluted boundary river dividing Moldavia from Roumania.

In February 1989 the Greens' efforts to hold an inaugural conference were supported by the Moldavian Writers, Journalists and Cinematographers' unions, but were opposed by local officials. Representatives from the Party Central Committee Secretariat and the Kishinev city soviet informed assembled delegates that their meeting was illegal, indicating that changes in the steering committee and formal association with the republic's nature preservation society would be needed to obtain official recognition.[14]

Rallies and demonstrations by unofficial Moldavian groups throughout 1989, combined with their successes in elections to the Congress of People's Deputies, underscored popular dissatisfaction over the environment, the language issue, élite privileges, consumer goods shortages and the republic's conservative leadership. By the middle of 1990, political power in Moldavia had shifted from the Communist Party to nationalist Moldavian forces. In the February–March republic elections, the Moldavian Popular Front and affiliated groups, including the Ecological Movement, gained control of the Supreme Soviet, and Moldavia subsequently declared its sovereignty from Moscow.[15]

Environmental problems in the Soviet Ukraine are extremely serious. They include contamination of at least four northern oblasts by radiation from the Chernobyl disaster; excessive use of toxic pesticides and herbicides in agriculture; soil salinisation from excessive irrigation; and a massive air and water pollution problem from the Ukraine's numerous mining and metallurgy enterprises.[16] Zaparozhe suffers from acute air quality problems; over 400,000 tons of toxic substances are released into the city's atmosphere each year. The situation is so critical that Moscow ordered a halt to all industrial construction in Zaparozhe in October 1989. Chemical pollution in Cherkassy has caused markedly increased rates of respiratory infections in children, a situation that prompted the town's nature society to collect 9,000 signatures on a petition protesting the chemical industry's disregard for public health.[17] The industrial centres of Dneprozherzhinsk, Dnepropetrovsk, Krivoi Rog, and the Donets Basin are all heavily polluted. In the western Ukrainian city of Chernovtsy, exposure to thallium, a poisonous metallic element, caused frightening hair losses in children in early 1989.

The water situation in the Ukraine is especially serious. Both the huge Dnestr and Dnepr rivers and many smaller bodies of water have been severely contaminated by agricultural run-off. Controversy has arisen over various irrigation projects in the region, most notably the Danube–

Dnepr and North Crimean canals, which have contributed to soil salinisation and erosion and exacerbate pollution problems of the larger Ukrainian rivers. Plans to construct a high-capacity pesticide plant near Shumskoye in the western Ukraine (30 kilometres from the Khmel'nitskii nuclear power station), a densely populated and highly fertile agricultural region, were cancelled following public protests.[18]

In February 1990 the Ukrainian parliament concluded that the republic was on the brink of an ecological disaster, resolved that the Chernobyl nuclear power station must be closed down, and announced plans to halt the discharge of all polluted effluent into bodies of water by the year 2000. Moscow received its share of the blame for these deplorable conditions: central economic departments were criticised for exercising a 'diktat' over siting and expansion of production facilities, for general economic mismanagement, and for 'ecological illiteracy'.[19] The Ukraine's July 1990 declaration of sovereignty conferred on the republic exclusive rights over all Ukrainian land and natural resources and established a Ukrainian committee for radiation protection. Parliament also averred its right to halt the construction or operation of industrial enterprises, including nuclear power facilities, that threatened the environment.[20]

In the Baltic states, where separatist sentiment is particularly strong, environmental issues have served as a rallying point for nationalism. Estonian environmental problems include degradation from phosphorite and oil shale mining, groundwater pollution from agricultural run-off, a rumoured nuclear waste dump at Paldiski, pollution of the Gulf of Parnu and Gulf of Finland, and fouled beaches on the Baltic Sea. Controversy over oil shale and phosphorite mining was instrumental in the formation of the Estonian People's Front and the Soviet Union's first Green movement which emerged in tandem during 1987. Since then, the Estonian Greens and the People's Front have cooperated closely in pressing nationalist and environmentalist demands on Moscow.[21]

In Latvia, the most serious environmental issues include pollution of the Baltic coastline by municipal sewage and factory wastes, the construction of a large hydro-electric station on the Daugava river and the impact of military operations on Latvian soil. The city of Riga, with a population of 1.2 million, lacks modern waste purification facilities; thousands of tons of wastes annually are dumped into the Gulf of Riga. Effluent from a pulp and paper mill at Jurmala has polluted the coastline, and offshore oil drilling by the Ministry of Petroleum further threatens the coastline and fishing stocks in the gulf.

Latvia's ecological movement originated out of a 1986–7 debate over the merits of a dam to be constructed on the Daugava River which would flood vast areas of forest and agricultural land and destroy many historical and cultural monuments. A protest effort led by writers, geographers and scientists resulted in cancellation of the project in July 1987.[22] Public protests, letter-writing campaigns and critical press coverage led the Latvian Supreme Soviet to order the Jurmala plant closed in early 1990, although it kept operating through the year on Moscow's orders.[23]

Much of the Baltic states' resentment against the central authorities has focused on Moscow's refusal to allow republican control over military recruitment and operations within territorial boundaries. The affront to Baltic sovereignty has been compounded by environmental consequences of military bases and manoeuvres in the republics. In Latvia, for example, the Supreme Soviet in August 1989 formed a working group to investigate violations of nature protection laws by military units, which included deforestation, improper disposal of wastes and the destruction of valuable agricultural land. The working group was directed to cooperate with local soviets and the Baltic Military District authorities to develop proposals for eliminating destructive environmental practices by the Soviet military.[24]

Russia and environmental politics

Seven decades of communist development policies have profoundly altered the Russian republic's natural environment. Russian nationalists greatly resent the environmental damage inflicted on their homeland, and most of the extraordinarily diverse currents of Russian nationalism accord a high priority to protecting the environment. The Bloc of Russian Public-Patriotic Movements, an umbrella organisation uniting twelve cultural, religious and political groups, incorporated environmental demands into its campaign platform for the 1990 elections to the Russian Republic Supreme Soviet. The Bloc's programme criticised central bureaucratic control over the republic's natural resources, disparaged official corruption and tendencies towards private enterprise as exploitative of Russia's natural wealth, and recommended the formation of a purely Russian set of administrative and economic institutions.[25]

Lake Baikal remains one of the most highly publicised examples of environmental pollution in the Russian republic. Continuing threats to Baikal's ecosystem from unrestrained development policies have mobilised Russian nationalists and ecologists. New revelations in the Gorbachev period confirmed suspicions that purification facilities mandated by the central government were inadequate to treat the effluent from the huge Baikalsk and Selenginsk cellulose plants. Reportedly, some twelve official decrees protecting Baikal were adopted during the Brezhnev period, yet none had any significant effect on improving the lake's purity. In addition to effluent from the pulp and paper mills, air pollution and additional wastes were contributed by the Ulan-Ude industrial complex in the Selenga basin, killing much of the zooplankton critical to the lake's natural purification system.

Public criticism of Baikal's pollution problems, deliberately suppressed in the later Brezhnev years, re-emerged in the 1980s. In the early Gorbachev period, deputies to the USSR Supreme Soviet and Russian nationalist literary figures mobilised public sentiment against the Ministry of Timber Industry. The first ecology group formed to protect the lake, the Baikal Movement, together with the Baikal Protection Society, organised protest demonstrations, sent letters of complaint and petitions to local Party and government officials, and successfully opposed a plan to divert the Baikalsk plant's wastes into the Irkut river. Their actions forced the Communist Party and government to decree a comprehensive clean-up of the region by 1995, involving conversion of the Baikalsk plant to a furniture factory and construction of a replacement pulp mill at Ust'-Ilimsk. A commission of academics, journalists and writers was created to monitor progress and to publicise ecological violations.[26]

Lake Baikal is a unique, even spiritual symbol for Russians and Buriats who inhabit Baikal's environs. Pollution of the lake has prompted students, scientists, politicians and workers to form alliances, and to plead for international support. In addition to the Baikal Movement and Baikal Protection Society, concerned residents have organised the Baikal National Front, the Centre for Ecological Defence of the Baikal Region, and Baikal Eco-World. The Baikal Fund has raised hundreds of thousands of roubles to help restore Baikal's former pristine condition.[27] No other ecological question, with the possible exception of massive radiation poisoning, can generate comparable emotions, and certainly no other ecological issue has a comparable effect on Russian nationalism.

Siberia and the Soviet Far East, the most sparsely populated regions of the USSR, experience some of the worst environmental problems. Geographical conditions – an extremely cold climate, the absence of hardy vegetation, a low regenerative capacity in rivers and lakes, and the fact that a substantial portion of Siberia is ecologically delicate tundra or permafrost – make this vast region particularly susceptible to environmental degradation. Of the thirty-six Russian Federation cities identified as having pollution levels four to forty-six times the permissible norms, nineteen are located east of the Ural mountains.[28]

In the Kuznetsk basin, industrial dust and sulphur dioxide air pollutants have raised the incidence of lung cancer, eye inflammation, and respiratory diseases. Novosibirsk, a relatively new city of 1.4 million people and a major steel and heavy machinery centre, is heavily polluted by sulphur compounds, nitrogen and particulates. Noril'sk, a small city far above the Arctic Circle, suffers from massive sulphur emissions from ore smelters. The giant Ob' river is polluted by oil wastes and spillage from the huge petroleum fields in western Siberia. Many Siberians are now demanding from Moscow greater authority to protect their land from careless exploitation.[29]

Lake Ladoga, the primary source of drinking water for Leningrad and the Karelian autonomous republic, has been severely polluted by several pulp and paper enterprises, by fertiliser and animal waste runoff from nearby farms, and by municipal sewage from the Leningrad community. To further complicate matters, Soviet planners decided to construct a system of dikes across the Gulf of Finland to protect Leningrad from periodic flooding. Many experts were convinced the dam would interfere with the natural sluicing of Leningrad's waterways, turning the canals into stagnant, open sewers.

The Russian intelligentsia raised a storm of criticism against developers' plans for the Gulf and against government inaction in dealing with Leningrad's water problems.[30] Criticism from nationalist writers, scientists, and Leningrad's ecology groups directed against the Gulf of Finland dam and pollution of Lake Ladoga eventually produced results. One major source of pollution located on Ladoga, the Priozersk cellulose plant, was closed down in 1987.[31] A commission of specialists from the Academy of Sciences and various ministries appointed in late 1988 to study the impact of the proposed dam described ecological conditions in Lake Ladoga, the Neva River, and the Neva inlet basin as abysmal, acknowledged that flood control measures would complicate the ecological situation, and proposed the formation of a permanent environmental oversight régime for the entire basin.[32] In August 1990 several prominent ecologists, including the deputy chairman of the USSR Supreme Soviet ecology committee Aleksei Yablokov and chairman of the group Ecology and the World Sergei Zalygin, charged in *Izvestiya* that the President of the USSR Academy of Sciences, Gennady Marchuk, was manipulating commission reports in order to convince the Council of Ministers to proceed with the project.[33] Finally, in October 1990, the Leningrad Soviet, on the recommendation of its ecology commission, suspended construction of the dam as ecologically and economically unsound.³⁴

Plans to divert water from the Ob' and Irtysh rivers of Siberia southward to irrigate Central Asian cotton fields provoked a major confrontation in the early and mid-1980s.³⁵ Opponents of the project included Russian nationalist writers, who argued that ancient Orthodox churches and other historical sites would be destroyed and the fragile Siberian ecosystem irrevocably transformed. Scientists expressed concern about the possible climatic effects of reducing water flow to the polar ice cap. Economists and geographers suggested it would be more cost-effective simply to improve efficiency in Central Asian irrigation; well over half of all water used for irrigation in the region was lost through evaporation or seepage from poorly-lined canals.

Promoting the river diversion scheme was a coalition of Communist Party and government officials in the Central Asian republics, scientists involved in the project, the USSR and Russian republic ministries of Land Reclamation and Water Resources who had invested heavily in preparatory work, some Gosplan officials, and representatives of the Central Asian intellectual community. Proponents argued that the diversion projects were necessary to meet demands for irrigation, for clean drinking water, and to replenish the shrinking Aral Sea. Prospects for future development in Central Asia, the poorest region of the USSR, would be bleak were the project to be cancelled.³⁶

By the twenty-seventh CPSU Congress, in February–March 1986, an official of the State Planning Committee announced that work on the canals was being suspended. A joint resolution of the Party Central Committee and Council of Ministers formally halted preparatory and design work in August, citing serious economic and ecological reservations, and directed Central Asian officials to improve water use efficiency by 15–20 per cent.³⁷ Debate over the merits of the projects continued in the pages of *Pravda*, *Nash sovremennik*, a conservative Russian nationalist literary monthly, and *Zvezda vostoka*, the Russian-language journal of the Uzbek writers' union.³⁸ However, the costliness of these diversion projects, together with the strength of opposition mounted by ecological forces in the Russian republic, make it unlikely that construction will resume in the foreseeable future.

Political participation and the environment

Environmental protection has figured prominently among the diverse causes advocated by newly-enfranchised Soviet citizens. Grass-roots environmental movements critical of the Communist Party and government's unwillingness or inability to deal effectively with diverse ecological problems have attracted supporters from virtually every national group and socio-economic category. Ecological issues have served as focal points for voluntary political participation, reflecting concern primarily over the health implications of a heavily polluted environment. An abysmal environmental record helped undermine the legitimacy of the old system and has strengthened calls for devolution of substantive decision-making powers to the local and regional levels.

Environmental politics has undergone a fundamental transformation from the Brezhnev to the Gorbachev eras. Environmental policy under Brezhnev, I have argued elsewhere, resembled a state corporatist model of participation. The Soviet state played the dominant role in recognising problems, placing issues on the public agenda, and modifying policies. Group activity, particularly among specialists, was an important aspect of Soviet environmental protection, but this participation was habitually manipulated and channelled to conform with régime priorities.[39]

As political liberalisation progressed in the Gorbachev period, independent ecological groups analogous to Western environmental movements began to emerge. Although many ecology movements encountered initial resistance from conservative local officials, hundreds or even thousands were estimated to have been formed by late 1990. Most of these groups were organised along republican lines and have frequently coordinated their activities with national front organisations. Other ecology movements are based in cities or universities or are dedicated to specific environmental causes. Some Moscow-based groups, most notably the Social-Ecological Union, function as umbrella organisations.

Pollution issues have generated extraordinary interest and activism among the Soviet population because of public concern over the health effects of environmental pollution. In a 1989 survey by the All-Union Centre for the Study of Public Opinion, 52.4 per cent of respondents listed health effects as the primary source of concern about the environment; the next closest response category (interfering with the natural order) was identified by only 13.9 per cent.[40] Pollution threatens not only the well-being of the individual; it raises the prospect of genetic mutation of one's immediate offspring and calls into question the long-term viability of society or a particular ethnic group. The fact that pregnant women and children often suffer disproportionately from pollution invests a highly emotional content into the issue. Soviet news reports have described in lurid detail the health effects of radiation poisoning, chemical smog, water pollution, and overuse of pesticides and herbicides in agriculture.

By rejecting obsessive secrecy for a more open approach to societal problems, glasnost has exacerbated people's fears about the possible consequences of pollution. The population has gradually become aware of the extent to which environmental problems were callously disregarded or covered up by authorities; however, suspicion remains that central ministries continue routinely to withhold information from the public. Chernobyl accelerated the shift to a more open society, yet continued reluctance on the part of Soviet officials to release full and accurate information about contamination levels in the Ukraine and Belorussia has further undermined public confidence.[41]

Environmental degradation has stimulated participation because it emphasises the lack of control individuals have over their lives in the Soviet system. As democratisation progressed under Gorbachev's leadership, Soviet citizens began to realise previously unavailable opportunities to limit or reverse government decisions that adversely affect their lives. Fear of toxic poisons and hazardous substances in the air, water, and food has generated a NIMBY ('Not in My Back Yard') attitude toward economic development issues. This increasingly vocal form of political participation is essentially negative, constituting a reaction to and opposition against the existing system. A NIMBY perspective will halt or slow environmentally questionable policies but will also complicate plans for economic reform.

Nationalism and environmentalism share two important factors which significantly affect political participation in the USSR. Firstly, both are emotionally charged issues. We have already noted the high level of concern about the medical consequences of environmental pollution among the Soviet population, which lends urgency to Soviet environmental activism. Nationalism is without a doubt one of the prime sources of identity for individuals in the Soviet Union today and therefore one of the strongest motivating forces for advancing political demands. Both nationalism and environmentalism have generated highly emotional forms of mass political participation that are directed against the political establishment.

Secondly, both environmentalism and nationalism are closely associated with physical territory. Although the identity of a national or ethnic group may exist and even thrive in the absence of a geographic homeland, the territorial linkage is central to the concepts of sovereignty, independence and self-rule – that is, the ability to define participation and decision-making rules free from external interference. Environmental problems perceived as imposed from outside by political forces of questionable legitimacy threaten the well-being of the territorial homeland and therefore strike at an essential aspect of sovereignty.

Conclusion

An examination of political participation, nationalism and environmental politics reinforces claims that we are witnessing the evolution of a 'civil society' in the Soviet Union.[42] Concepts of limited government, rule of law, civil rights liberties and the responsiveness of government to public opinion are taking hold. However, the form of participation that is emerging is somewhat different from that usually found in established participatory democracies. Political participation in much of the USSR is populistic, embodying a distrust of the rational and calculating aspects of modern society and affirming a belief in the creativity and moral superiority of ordinary people.

The populist essence of current Soviet politics relates to the breakdown of mass organisations, which served, however imperfectly, as mobilisers linking élites and masses in the totalitarian structure. During this period of rapid transition, Soviet society has exhibited symptoms of anomie – old values, norms and patterns of behaviour have broken down, leading to the erosion of political legitimacy.[43] It is under conditions of anomie and atomisation that individuals are most susceptible to populist-style manipulation directed against a corrupt or unresponsive entrenched élite. Urban populism in particular appears to be connected to a crisis of the dominant ideological paradigm arising from frustrated expectations.[44]

The ideologies that populism relies on are subtly or overtly nationalistic or even racist, providing a sense of belonging frequently lost in a modernising society. Although the environmental attitudes of most Soviet citizens are far too narrowly based to provide such an ideological framework, political activists can combine ecology demands with nationalist issues to create an agenda with strong populist appeal. Green politics, frequently praised as direct democracy unmediated through hierarchical organisations, dovetails easily with the populist politics of nationalism. Sadly, instead of providing a cohesive value structure that could help reintegrate society, environmentalism appears, if anything, to reinforce the disintegrative forces of nationalism driving Soviet politics.

Notes

1. This paper draws in part on the author's 'Environmental politics and policy under Gorbachev', in *Unofficial Movements in the USSR*, edited by Judith Sedaitis and Jim Butterfield (Boulder, CO: Westview Press, 1991). An earlier version was presented to the Fourth World Congress of Soviet and East European Studies, Harrogate, England, 21–26 July 1990.
2. Important contributions to the debate on participation in the USSR include Theodore H. Friedgut, *Political Participation in the USSR* (Princeton: Princeton University Press, 1979); Jerry F. Hough, *The Soviet Union and Social Science Theory* (Cambridge, MA: Harvard University Press, 1977); Everett M. Jacobs (ed.), *Soviet Local Politics and Government* (London: Allen and Unwin, 1983); Jeffrey W. Hahn, *Soviet Grassroots: Citizen Participation in Soviet Local Government* (Princeton: Princeton University Press, 1988).
3. See Philip P. Micklin, 'Dessication of the Aral Sea: a water management disaster in the Soviet Union', *Science*, 241, no. 4870 (1988), pp. 1170–6; and William S. Ellis, 'A Soviet sea lies dying', *National Geographic*, 177, no. 2 (February 1990), pp. 7–92.
4. The figure for the entire USSR is 8.4 per cent. A. Vasil'ev and M. Krans, 'Aral: varianty reshenii', *Kommunist*, 2 (1990), p. 56.
5. Associated Press, 16 June 1989.
6. See Bess Brown, 'The role of public groups in *Perestroika* in Central Asia', Radio Liberty *Report on the USSR*, 2, no. 4 (26 January 1990).
7. Vasil'ev and Krans, 'Aral', pp. 55–65.
8. The best recent review of Armenia's ecological situation is A. L. Valesyan, 'Environmental problems in the Yerevan region', *Soviet Geography*, 31, no. 10 (October 1990), pp. 573–86.
9. Elizabeth Fuller and Mirza Mikaeli, 'Azerbaijan belatedly discovers environmental pollution', *Radio Liberty Research* (RL 2/88), 30 December 1987.
10. Yasin Aslan and Elizabeth Fuller, 'Azerbaijani press discusses link between ecological problems and health defects', Radio Liberty *Report on the USSR*, 1, no. 31 (4 August 1989).
11. Mirza Mikaeli and William Reese, 'The Popular Front in Azerbaijan and its program', Radio Liberty *Report on the USSR*, 1, no. 34 (25 August 1989).
12. Zeev Wolfson, '"Nitrates" – a new problem for the Soviet consumer', Radio Liberty *Report on the USSR*, 1, no. 20 (19 May 1989).
13. Grigore Singurel, 'Moldavia on the barricades of *Perestroika*', Radio Liberty *Report on the USSR*, 1, no. 8 (24 February 1989).
14. Vladimir Socor, 'The Moldavian Greens: an independent ecological association', Radio Liberty *Report on the USSR*, 1, no. 11 (17 March 1989).
15. See Vladimir Socor, 'Moldavia: political power passes to democratic forces', Radio Liberty *Report on the USSR*, 3, no. 1 (4 January 1991).
16. For more complete information on Ukrainian ecology problems, see the

various reports by David Marples in Radio Liberty's *Report on the USSR* for 1989–1990.
17. Judith Pereira, 'Where glasnost meets the Greens', *New Scientist*, 1633 (10 October 1988), pp. 25–6.
18. TASS, 22 February 1990.
19. *Pravda*, 18 February 1990, p. 3.
20. Peter Shutak, 'Ukraine declares sovereignty', *Soviet Analyst*, 19, no. 15 (1 August 1990).
21. See Dimitri Devyatkin, 'Report from Estonia: an interview with a leader of the Green Movement', *Environment*, 30, no. 10 (December 1988), pp. 13–15.
22. Nils R. Muiznieks, 'The Daugavpils hydro station and "glasnost" in Latvia', *Journal of Baltic Studies*, 18, no. 1 (Spring 1989), pp. 63–70.
23. Dzintra Bungs, 'Latvia: high hopes and harsh realities', *Radio Liberty Report on the USSR*, 3, no. 1 (4 January 1991).
24. *Foreign Broadcast Information Service: Soviet Union*, 16 February 1990.
25. *Literaturnaya Rossiya*, 29 December 1989, translated in *Current Digest of the Soviet Press*, 42, no. 1 (7 February 1990), pp. 1–4, 23.
26. See Zeev Wolfson, 'Anarchy mirrored in Lake Baikal', *Radio Liberty Report on the USSR*, 1, no. 21 (26 May 1989).
27. See John Massey Stewart, '"The Great Lake is in great peril"', *New Scientist* (June 30, 1990), pp. 58–62.
28. *Pravda*, 1 September 1989, p. 8.
29. For an informative discussion of Siberia's ecological problems, see Mike Edwards, 'Siberia: in from the cold', *National Geographic* (March 1990), pp. 2–39.
30. Yurii Bondarev, speaking at the 1986 congress of Soviet writers, equated the protection of Lake Ladoga, and the natural environment more broadly, with the protection and preservation of Russian culture. *Literaturnaya gazeta*, 27 (2 July 1986), p. 10.
31. *Sovetskaya Rossiya*, 29 May 1987, p. 1.
32. *Leningradskaya pravda*, 21 April 1989, translated in *Current Digest of the Soviet Press*, 41, no. 19 (7 June 1989), pp. 23–4.
33. *Izvestiya*, 17 October 1990, p. 2.
34. *Izvestiya*, 17 October 1990, p. 2.
35. See Philip P. Micklin, 'The vast diversion of Soviet rivers', *Environment*, 27, no. 3 (March 1985), pp. 12–20, 40–5.
36. A debate over the merits of the diversion project was conducted in the pages of *Novyi mir*: See Sergei Zalygin, 'Povorot: uroki odnoi diskussii', *Novyi mir*, 1 (January 1987), pp. 3–18; and 'Kak sovershaetsiya povorot', *Novyi mir*, 7 (July 1987), pp. 181–235. Also, see Bess Brown, 'What will cancellation of the Siberian River Diversion Project mean to Central Asia', *Radio Liberty Research* (RL 334/86), 26 August 1986.
37. *Pravda*, 20 August 1986, p. 1.
38. Peter Sinnott, 'Water diversion politics', *Radio Liberty Research Bulletin* (RL 374/88), 17 August 1988.
39. Charles E. Ziegler, *Environmental Policy in the USSR* (Amherst: University of Massachusetts Press, 1987), pp. 45–77.

40. *Ogonyok*, 50 (December 1989), p. 3.
41. See Zhores A. Medvedev, 'The environmental destruction of the Soviet Union', *The Ecologist*, 20, no. 1 (January–February 1990), pp. 24–9.
42. S. Frederick Starr, 'Soviet Union: a civil society', *Foreign Policy*, 70 (Spring 1988), pp. 26–41.
43. See Elizabeth Teague, 'The Soviet "disunion": anomie and suicide', Radio Liberty *Report on the USSR*, 2, no. 47 (23 November 1990).
44. Ernesto LaClau, *Politics and Ideology in Marxist Theory* (London: NLB, 1977), pp. 151–2, 175.

4 BAM after the fanfare: the unbearable ecumene

Victor L. Mote

Introduction

> God be with you, my friends,
>   in storms, in your worldly sorrow,
>   in that netherworld, that desert-like sea ...[1]
>   Mikhail Glinka on the Decembrists in Siberia

The aim of this chapter is to describe the contemporary physical and human ecological conditions inherent in the service area of the Baikal–Amur Mainline (BAM) railway and to explain in part the reasons for those conditions. The geographical concept of ecumene is employed to satisfy this purpose.

An ecumene is a geographic notion used to describe a permanently inhabited area of the earth.[2] It is a didactic tool employed to illustrate the difference between those parts of the earth that are suitable for human habitation and those that are not. Thus, geographers can calculate a more realistic approximation of population density by relating the size of a population in a given region to the size of its ecumene or livable space, disregarding all the excess that is uninhabitable.

Glinka was not alone in his impression that Siberia was uninhabitable. For generations, tsars viewed the region as a dismal land to which to send criminals, recalcitrants, potential and real revolutionaries, and many other poor souls falsely accused of even the simplest of crimes.[3] Between 1823 and 1887 some 772,979 persons were exiled to different parts of Siberia. According to Treadgold,[4] migrant exiles exceeded peasant migrants by a factor of 1.7 between 1800 and 1880, after which the government finally encouraged free peasant settlement of Siberia on a considerable scale. Burgeoning first in the steppes and wooded steppes of Western Siberia, a steadily increasing stream of settlers pushed the frontier eastward all the way to the Pacific.[5] Between 1880 and 1914, 4.3 million free migrants arrived in Siberia, outnumbering exiles by a factor of more than 4 to 1.[6]

Virtually all of this process of settlement stayed within the more hospitable southern tier of Siberia and what is now the Soviet Far East. Although frigid exclaves like Turukhansk (1607), Yakutsk (1632), and Anadyr (1649) were settled early on,[7] they remained, and still remain, largely isolated outposts that are well beyond the common understanding of what is, and is not, ecumene. Indeed, the most dedicated efforts to push the tier of settlement into the so-called 'Near' and 'Far North' of Siberia have come during the Soviet period.[8] The most outstanding examples of this recent 'push to the north' must be those associated with the development of Western Siberian oil and gas and the construction of the BAM railway in Eastern Siberia and the Soviet Far East.

This chapter focuses on the BAM and its tributary area since its rails were linked in 1984. Until that year, the BAM and its construction heroes were the subjects of almost never-ending praise. Under Leonid Brezhnev (1964–82), who took more than a personal interest in the project, the needs of the BAM and its workers were given the highest priority. Under Gorbachev, who took over in 1985, new priorities were set and BAM was de-emphasised to the extent that it is now an object of scorn. The de-emphasis has not failed to affect the lives of tens of thousands of contemporary settlers in the BAM region. Indeed, the first half of the title of this chapter is derived from a popular Soviet television documentary aired first in 1988, which revealed the BAM and its infrastructure in graphic detail.[9] The reality of BAM is that it has engendered an ecological nightmare in both physical and human aspects. Straining the limits of human settlement, it has forced humankind into Glinka's 'netherworld' and 'desert-like seas'.

BAM and physical ecology

The physical consequences of BAM were easily predictable (figure 4.1). From its inception, it heralded dramatic environmental impact. When its 3,500 kilometres (BAM plus 'Little BAM') were linked in 1984, Soviet news media vaunted the following successes: 9 tunnels with a total length of 30 kilometres; 3,136 water crossings, 142 of which were longer than 100 metres; over 200 railway stations, 50 settlements, and dozens of sidings; thousands of kilometres of frontage road; and 220,000 cubic kilometres of earthwork (5 times greater than expected in 1975).[10] The earthwork alone represented a volume that could inter 95 per cent of England, Scotland, and Wales beneath a pile of overburden three-quarters the height of Ben Nevis (1,100 metres). In the United States, debris of the same volume and thickness could bury the entire state of Utah.

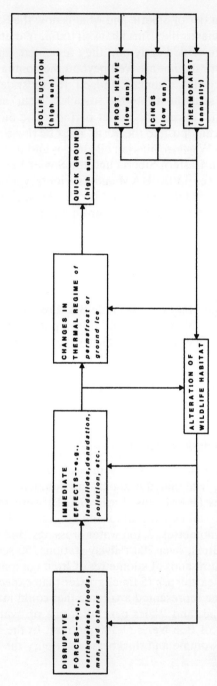

Figure 4.1 A model of perpetual disruption in a periglacial environment

Permafrost problems

In periglacial environments such as the BAM zone, the arch nemesis of the engineer is permafrost. In such environments, only the upper few feet of the surface (active layer) freeze and thaw seasonally. The remainder of the ground stays solidly frozen throughout the year; hence, the name 'permafrost'. This yearly cycle of freezing and thawing induces a surface rhapsody known as 'annual pulsation', which eventually yields a phenomenon called frost heave that thrusts anything in the active layer surface-ward. Permafrost is affected most seriously by surface disturbances, especially the wanton destruction of the vegetative layer, which accelerates temperature changes and associated melting. These lead inexorably to differential settlement and subsidence, the ultimate result of which can be wholesale disruption on a regional scale, that which geomorphologists call 'thermokarst' (karst topography induced by melting in place of the corrosion of limestone). This not only changes the appearance of the original landscape, it can also affect elevation and drainage patterns.

For two-thirds of its length, BAM lies on permafrost. From their inception, the railway's subgrade and foundation have been attacked by the fore-mentioned processes, which have included both frost heave and permafrost melt. To counter these problems alone, the government has had to earmark an annual subsidy of ten million roubles per year ($13 million 1988 US).[11] However, associated with these basic periglacial problems have been outbursts of 'supercooled' groundwater known as *taliki* that, once exposed to the ambient air, freeze into 'icings'. Such outbursts can have an almost 'volcanic' impact, destroying or inundating structures or road-beds within range of the phenomenon.

Many of BAM's problems with permafrost might have been avoided had Moscow central planners and engineers paid heed to the advice of local specialists based in Eastern Siberia and the Soviet Far East, some of whom had had experience with building railways on permafrost as far back as 1927. However, the central authorities chose not to do that. Far eastern experts argued for a subgrade of appropriately-sized ballast that would be allowed to force-freeze and become an organic part of the permafrost before allowing further construction. Their arguments were overruled by authorities at Gosplan, who complained that construction would have been delayed by the technique. According to the far easterners,

It was difficult to countermand the authority of the venerable specialists in the capital. The majority of them were steadfast proponents of the traditional concept that allowed for some deformation of the railway bed . . . and they simply refused to listen to any other opinion. They even threatened to take us to court. This is how our permafrost experts were harried by highly placed authorities who were willing to sacrifice science to the satisfaction of their own personal ambitions.[12]

Considering any disruption of the permafrost anathema, the far eastern experts bore subsequent witness to the chain reactions displayed in figure 4.1 above. Not only have they seen permafrost damage, but they have also observed river bank erosion and enhanced stream turbidity, especially in the regions of Urgal, Fevral'sk, and the Muya Lowland.

Again counter to the recommendations of the local experts, central planners allowed workers to divert the Urgal River some seven kilometres around the site of a new town of the same name. The canal that was dug naturally disrupted the permafrost, changed the water table, and destroyed 60 hectares (150 acres) of forest. The thermokarst development that has ensued will expand because, according to observers, *talik* water under the railway foundation has been redistributed.[13]

Because planners chose haste over waste during BAM's construction, the railway was laid through the site of the now existing town of Fevral'sk, which was not included in the original plan. Fevral'sk is underlain by sporadic permafrost, which was not detected by the designers of the revised route. Solid ground exists a few kilometres west of the existing line and would have served as a better foundation. Contemporary Fevral'sk, which lies on the left bank of the Selemdzha River, is described as a city in the midst of a swamp. Differential settlement caused by the melting of the sporadic permafrost has broken residential foundations, ruptured sewage pipes, cracked masonry, and induced other undesirable consequences. The Selemdzha itself, which at one time was assessed as an excellent recreational river, was despoiled to a great degree by quarrying and excavating of gravels for BAM railway ballast.[14]

Activities associated with the boring of the Muya Tunnel, including the tunnel itself and two tunnel bypasses (one a failure and the other now operating), have disrupted one of the BAM's most unique ecosystems. Excavation and other earthwork have altered the natural drainage patterns, lowered the water table, dried up several marshes, dessicated cedar thickets and copses of the rare Muya rhododendron, and killed aquatic life because of increased stream turbidity. Intense logging of the hillsides surrounding the Muya Valley has melted much of the permafrost and accelerated the run-off to the extent that in the estimation of biologists there may be no possibility of reforestation.[15]

Physiography

For much of its length the BAM railway is rarely far from seismically active areas. Its western and central zones pass through deep grabens of the Mongolian-Baikal rift. Earthquakes measuring 4 or higher on the Richter scale have been recorded between Ust'-Kut and Tynda, the highest being 7.9 (in the Muya and Chara valleys in 1957).[16] The earthquakes have triggered some of nature's worst examples of mass wasting: snow and rock avalanches, mudslides, and mudflows.

The settlements of the BAM service area are, naturally, products of the railway. Railroads, to perform normally, follow paths with slopes of less than 5 per cent; thus, BAM cities and towns are found in valleys. In the western BAM, the valleys are steep-walled, flat-bottomed, normally faulted grabens that are subject to intense and persistent temperature inversions. (The world's highest air pressures are recorded in Eastern Siberia: almost 1080 millibars.) Air drainage occurs almost every night of the year and is especially severe in the winter because of the persistence of the 'Siberian high' (the world's 'hyperbar'). Because of the accompanying atmospheric stagnation, parts of Chita oblast experience air pollution that is on average eight times worse than Soviet allowable norms, even in towns without a single industrial enterprise.

BAM and human ecology

Because of its uniqueness and concomitant sensitivity, Lake Baikal is in a state of perpetual ecological crisis owing to human activities within the lake's watershed. Even before the uproar over the Baikal'sk Pulp and Paper mill reached the headlines of Soviet newspapers in the 1960s, Baikal's shoreline was being systematically denuded of forests. Although this activity allegedly had been banned by decrees of both the Party and government, renewed clearing along the north shore associated with the construction of BAM became known in the 1970s.

In 1978 geographer L. N. Il'ina championed the cause of the lake and the whole BAM zone ecology when she argued persuasively for the total preservation of Baikal's shoreline forests, selective logging in south Yakutia, and the careful clearing of trees along the entire length of the BAM railway. The Forest Ministry disagreed and Il'ina's recommendations were tabled until 1986.[17] Under Gorbachev, she was encouraged to resume her ecological studies on site, this time with a full staff from the Moscow-based Institute of Geography. Their efforts have yielded a detailed map of BAM ecology with which they and others can monitor environmental problems in the service area. Il'ina and her staff have called for, among other things, forest-protection belts of at least 500 metres in width along rivers of medium size and larger and of 200 to 300 metres along smaller water courses. Not only do such belts protect the permafrost in areas where they exist but they also retard run-off, reduce erosion, and lower stream turbidity.

Il'ina's efforts have had practical political impact. The Siberian 'Greens' have coopted some of her scientific studies into their arguments with local authorities. In response to their advocacy, the Irkutsk *oblispolkom* in late 1989 repudiated proposals for five new timber-industrial complexes and railed against clear-cutting along the route of the BAM.[18] Moreover, in late December 1990, a new ecological service based in Tynda and affiliated with the State Agency for Railway Construction was given jurisdiction over the entire BAM service area. According to its Chief, T. I. Zimnina, who is no doubt in contact with Il'ina's team, all BAM area enterprises are to be given 'ecological permits' in order to operate within the tributary area. Once in operation, the enterprises will be computer monitored by Zimnina's ecological service in the BAM capital. In contrast to past enforcement, stiff fines will be levied on violators of the ecological laws, which are based on standards peculiar to the BAM service area. According to Zimnina, enterprise managers will find it far cheaper to protect the environment than to disrupt it.[19]

The other side of Lake Baikal

Il'ina believes the most serious ecological problems in the BAM zone are those associated with Lake Baikal. The notoriety gained by Baikal'sk during three decades of controversy has overshadowed other deleterious developments along the lakeshore and in its watershed. Il'ina's team has determined that there are at least 150 such developments, not the least being those associated with the troubled city of Severobaikal'sk.[20]

Severobaikal'sk has had a tumultuous history. By 1990 the city claimed 34,000 residents, even though it realistically could support only 14,000. The statistical paradox stemmed from a past record of poor planning: the original (1975) plan called for 25,000 residents, cut back to 14,000 in the late 1970s, but raised again to an overly optimistic 140,000 in the heady early 1980s. The last blueprint envisioned a city with two large machine-building factories. Under Gorbachev, these have not been, nor will they be, built. Meanwhile, anticipating the new industrial development, 20,000 excess migrants were drawn to the beautiful setting on Lake Baikal's northern shore. Consequently, three out of four of Severobaikal'sk's citizens live in sub-standard housing that includes among other things trailers and unserviceable rolling stock. Most of these dwellings are without indoor plumbing and running water. Residential areas are serviced by a system of twenty-six neighbourhood ('pygmy') boilers, only one of which is equipped with any kind of air pollution abatement (in this case ash collection filters). In 1988 an estimated 18,000 tons of suspended particulate was emitted from the smokestacks and chimneys of Severobaikal'sk.[21] The city was also shrouded by a gaseous haze of sulphur and nitrogen compounds, the principal components in acid rain. Local forests have begun to change colour and die. This has enhanced surface run-off, which, of course, flows into Lake Baikal via the Tyya and Lower Angara rivers. The human-tainted precipitation that raises the level of the rivers is steadily augmented by petroleum products from local motor pools and heavy metals from all the new construction. Fish kills, which include the famous Baikal omul, have occurred in the Lower Angara basin near the town of Kichera.[22]

Lake Baikal's northern shore is easily as polluted as the southern shoreline near the Baikal'sk mill. After all, the Baikal'sk Pulp and Paper mill was equipped with a three-stage water pollution abatement system and still had problems. Tertiary systems such as those at Baikal'sk are the most expensive in the world. But the by-products being generated in Severobaikal'sk and other BAM settlements in the Lake Baikal watershed are released without even the most primitive pollution control; thus, in 1988, Il'ina's team of geographers discovered 'dozens of harmful substances' in the waters off Severobaikal'sk.[23] The inventory undoubtedly has grown since then. Perhaps even worse, pirating of the forests along the north lakeshore and the creeks and rivers that feed the lake persists, leading to erosion and stream turbidity. Ecologists estimate that the annual cost of environmental deterioration in the north Baikal region has now surpassed 20 million roubles.[24]

Udokan: a would-have-been Chernobyl?

Severobaikal'sk is certainly not the only ecological tragedy among the BAM settlements. The would-be copper-mining centre of Udokan also has a controversial history. A decade before renewed construction on the east–west axis of the BAM railway was announced, the first stage of exploration of the Udokan copper basin was completed. The Udokan site, which was discovered in 1948, is reputed to be the world's largest copper region, but its ores have yet to be processed commercially because they are complexly polymetallic and, without BAM, have been too remote.[25] Soviet authorities have tried desperately, to no avail, to interest foreign companies (mainly Japanese) in developing the site.

In the 1960s, however, they were content to try to develop the site on their own. The draft of the ninth five-year plan authorized the Udokan programme, but did not go into detail. What the State secretly authorised was the use of a peaceful nuclear explosion on or near the surface of 'drift mine No. 5'.[26] The explosives actually were delivered by Trans-Siberian railway to Mogocha station and carried over the primitive winter road to Naminga during the winter of 1965–6. The Chara River, meanwhile, burst forth from its frozen surface and created a lateral river icing, which delayed the shipment. When the demolitions finally were delivered in late 1965, local geologists were evacuated beyond a 50-kilometre radius of ground zero. An Evenk village was moved, but authorities in Moscow decided not to evacuate the larger settlement of Chara for fear that it would cause more harm from panic than would result from the potential exposure to radiation.

Drift mine No. 5, it turns out, was the site of the poorest grade of ore, but if the experiment had succeeded there – and no one ever doubted it would – they would have proceeded to the sites of the richer ores.[27]

This 'peaceful use of the atom' was planned for some time between January 2nd and 6th, 1966, a time when the winds were expected to be (and usually are) exceedingly low, 'but still they blow and do so unimpededly because there are no obstacles at Udokan's elevation'. As in many parts of the Udokan basin, the rock is highly fragmented and latter-day experts are convinced that the potential radiation would have reached the water table and eventually the Lena River system. Although the Udokan explosion was called off at the last moment, the Brezhnev administration had planned to cordon off the region for ten years in order 'to let the radiation return to normal'.

Although peaceful nuclear explosions for resource development have been used in the United States, since the 1950s they usually have been carried out far below the surface. Moreover, the public has been informed well in advance of where and when those explosions would occur. The Udokan experiment was to have been carried out in total secrecy and virtually above ground. Some twenty-five years later, its story finally broke to the Soviet public.

Although Udokan was never contaminated by nuclear radiation, the construction of the BAM has led to a number of untoward circumstances. In 1979, the then roadless town of Udokan was traversed by the only means of travel available: tracked vehicle. Although the sad results of the use of such vehicles on permafrost had been pictured prominently in Soviet earth science textbooks and monographs for several dozen years,[28] BAM workers used them almost indiscriminately in the Udokan region. The tracks of the use of those vehicles still remain there except that they now consist of gullies collectively the size of more than five contiguous American football fields almost 9 metres deep.[29]

Other BAM environmental problems

The scars visible at Udokan are alleged to appear in all the towns and cities of the BAM service area. The vast majority of the settlements were assembled rapidly on clearcut and bulldozed ground largely without consideration for the underlying permafrost or the sensitive soil and vegetation on top of it. Urban residents are reported to 'breathe dust in summer and soot in winter'. Conservation of the forests might have aided the citizenry by absorbing up to one-half of the natural dust and by-products of combustion. But, with most of the early settlers self-reliantly using forest products to build their accommodation and burning firewood as fuel, the treeline now begins on the edge of the towns. In the urban interiors, the landscaping that exists consists mostly of recent (post-1980) plantings. Because BAM forests take 100 years or more to mature, many years will be needed before the saplings will have an environmentally beneficial effect. As Gorbachev took office in 1985, virtually all of the housing and office space in Kirenga, Magistralnyi, Ul'kan, and Kunerma were made of wood. Cabins had ample wood piles for use in saunas and stoves. Only the larger steam and power stations burned coal.[30]

In the early 1970s, the BAM was envisioned as 'a super mainline' with all the attendant state-of-the-art technology. The towns and cities, stations and depots were mere afterthoughts. The principal aim was to build the railway as fast as possible. In the autumn of 1989, BAM's chief engineer V. A. Gorbunov complained that the settlements of the region had obsolete water systems, sewage treatment installations, and central heating plants.[31] Earlier, having examined forty-five BAM settlements, a high-ranking party control committee considered no more than twelve as 'suitable for human habitation'.[32]

Water supply and sewage treatment are especially difficult problems in periglacial environments. Soviet plumbing is notoriously bad, but in Siberia, particularly where permafrost is a factor, water and sewage pipes must be protected or heated. Where public utilities exist in the BAM service area, insulated pipe of large diameter is placed on pilings or gantries which rise gawkily over roadways and other thoroughfares. Evidently many BAM towns are still without running water or flush toilets. In December 1990, representatives to the RSFSR Supreme Soviet bemoaned the living conditions of the people who reside along the older 360-kilometre Urgal-Izvestkovaya rail spur, where 'dozens of families "camp out" in wooden barracks' that have no running water or conveniences. Ironically, these are the same barracks that were built by Japanese prisoners-of-war in the immediate post-war period (1945–7).[33] Earlier (1988), in the new town of Yuktali (formerly Ust'-Nyukzha), where utilities do exist, drinking water supplies were polluted by systematic leaks from broken sewer mains. Moreover, because the sewage was warmer than the permafrost on which it spread, it induced thermokarst processes that deformed the BAM railway embankment.[34]

The activity of decomposers is sharply curtailed in periglacial environments. The normal disintegration of rubbish or natural organic matter occurs at best in the brief summer months, and even then, only at high sun (11 am to 3 pm). Preserved below by permafrost and above by the lack of decomposers, rubbish accumulates in slowly decaying mounds or landfills, usually along the long-frozen rivers.

In Tynda, for instance, a city of 60,000 people, the municipal dump was located too close to the once-pristine Gilyui River. However, it was too far away for some rubbish collectors, who dumped their loads into the city's surrounding forests. During the summer of 1988, the Gilyui flooded, sweeping away all but the heaviest debris from both the dump and the forest. The solid waste eventually reached the Zeya River reservoir (250 km downstream). The stagnant reservoir, already suffering from the effects of slowly decaying forests flooded by the Zeya dam a dozen years before, thus became an 'ultimate sink' for BAM area rubbish.

Tynda, like Severobaikal'sk, is a city with special, if not serious, environmental problems of which solid waste disposal is only one. According to one report:

Tyndans have long ago forgotten what it's like to see white snow. Until recently, the city had around 100 pygmy boilers that belched out some 500 tons of ash and soot every year. Today [May 1990], many have been shut down and the situation is better, but in April, the local sanitary-epidemiological station reported that in the vicinity of the railway station, air pollution exceeded State norms by a factor of 24 and in one residential area by a factor of 22.[35]

The reasons for such high readings are obvious. Where they exist at all, to save money, central heat and power plants were built within the city limits of the BAM settlements, usually near the city centre close to the railway station. As the towns expanded into true cities, the power plants became even more central. Tynda's main plant is *upwind* from a major residential zone within the same valley. The many other steam boilers, all of which are fuelled by steam coal from Neryungri, only add to the problem.

Neryungri steam coal consists of detritus from the world's largest coal washery at Ugol'naya 222 kilometres north of Tynda. It is the same type of coal that has been burned for over forty years in a power plant in the coal-mining town of Chul'man, 38 kilometres further north. The coal is high in ash (9 to 10.2 per cent), which explains why Tynda and other central BAM settlements are ringed by ash heaps, the particulate from which 'blows all over the place' on windy days.[36]

According to the same observer, who proposes a 'Novgorodian solution':

> It's not just Tynda: everywhere you find a BAM settlement, it's the same – smoke, smoke, smoke! BAM needs to be set up as an 'Open Economic Zone' so that the Chinese or Japanese can come and clean up our mess![37]

Indeed, in 1985, a foreign visitor to Chul'man was horrified to see black sooty snow. Soot covered everything and everyone in sight.[38] There was nothing 'white' about Chul'man, which means 'white rock' in Evenk.

On the plus side Neryungri steam coal is low in sulphur (0.3–0.4 per cent) and phosphorous (0.004 per cent). In July 1990, the ash residue was determined to be rich in titanium, magnesium, aluminium, and flint, all of which have so far ended up in the local ash heaps, polluting the environment.[39] BAM workers have been advised to use the residue as fill or railway ballast, and some is converted on site into cinder block for use throughout the service area.

Despite automobile prices 10–15,000 roubles higher than in the centre, cars (and car exhaust fumes) have also increased. Tynda has one out of every five private automobiles in the BAM region (5,000 out of 25,000) and only three car repair shops (Soviet standards recommend a minimum of eight for such a fleet). The owners of the 20,000 other cars in the service area are forced to wend their way over mostly dirt roads to Tynda to have their vehicles repaired or serviced. What this means environmentally is that many cars in the BAM zone doubtless operate untuned and thus pollute more severely. Tynda authorities, employing Gorbachev's 'new economic mechanism', have tried to entice new garage owners to the BAM, but no one has been interested. 'The costs outweigh the benefits', say officials at the Volga automobile plant.[40]

Conclusion: is there life on BAM?

Ecumenes are simply zones that have proved to be habitable; there is no qualitative context implied by the definition. One of the best indicators of the quality of an environment in an industrialised society is the quality of life of its people. It has been often said that environmental concern rises concomitantly with per capita real income (and leisure) levels. In other words, people must be rich enough in an industrialised society to pay for the dis-economies generated by production processes. When an industrial economy is worked by people who are more concerned about food and shelter than they are about the environment, the environment usually is of lower priority.

Environment is a comprehensive term that includes both human and physical factors, the equivalent of the French notion of 'milieu'. Accordingly, environmental quality bespeaks the quality of both the physical and human environment. Where standard of living is concerned, many incredibly poor people have lived (and still live) in beautiful, pristine environments, whereas some of the world's wealthiest people have been surrounded by often insalubrious conditions. The worst condition arises when poor people must reside in unwholesome environmental conditions.

In the heady days just after Yury Gagarin led the Soviet people into the age of manned spacecraft, Soviet citizens joked: 'Why should we want to go to Mars? There isn't any life on Mars either'. Since the glorious days when the last 'silver spike' was driven into the ultimate 'golden link' at the nondescript town of Kuanda, reality has descended on BAM. The region not only suffers from a seriously disturbed physical environment but under perestroika it has undergone an economic turn for the worst. BAM's most decorated construction hero, Alexander Bondar', for awhile found himself unemployed in his chosen residence of Severobaikal'sk. By the end of 1990, unofficial unemployment in Chita oblast had reached 9,000, two-thirds of whom were young people, and was expected to top 100,000 within one year.[41] Many of those unemployed were found in the BAM service area.

Shortages of foodstuffs were manifest all over the USSR during the winter of 1990–91, but they were *very* severe in the BAM service area. Ordinarily triple the price of goods found in 'southern regions' (like Blagoveshchensk), the cost of something as mundane as a jar of pickles jumped from 2 to 6 roubles in a 24-hour period on the shelves of BAM zone stores. Everything, 'even what was stale and rotten', was bought without question. Compared to the spartan year of 1989, BAM in 1990 experienced shortages of meat (1,800 tons), dairy products (8,000 tons), and eggs (5.5 million units). Forget about potatoes on BAM: 'It's difficult to raise potatoes on permafrost and feedstocks for livestock are nonexistent'. Natural whole milk for children could not be bought. Even bread was hard to find in some parts of the service area. The markets were said to have returned to 'the stage of feudalism'. In Tynda, the situation was reported to be 'explosive'.[42]

The acquisition of adequate shelter is particularly difficult in the settlements of the BAM service area. As of autumn 1990, 85 per cent of the BAM residents 'who would like to stay' in the zone were dissatisfied with their housing.[43] In a country with one of the lowest per capita housing allotments in Europe in 1988, the Russian republic averaged around 14 square metres (148 sq ft) per person, the Soviet Far East even less (12.7 sq m/134 sq ft), and the BAM half the latter.[44] One woman, living in Kirenga on the western BAM, related that she, her husband, and her daughter still shared 7 square metres (74 sq ft) of living space after twelve years of work on the railway.[45] That computes to less than 25 square feet per family member or half the size of an American prison cell. In a sample of 132,000 BAM families in 1988, only 24,000 were found to be living in housing built since 1978; the rest lived in 'temporary shelters' such as trailers, abandoned rail cars, or barracks.[46] By 1990, an average of seven to nine years were required to obtain a new apartment in the region.[47]

The BAM capital typifies the problems of housing and infrastructure that are rampant throughout the service area. In Tynda, 11,000 family names are on the waiting list for new apartments. Because some 500 to 600 new units are built there each year, those families at the end of the list will have to wait for more than twenty years to make their move! Many families in the city have young children: each year 1,500 babies are born, keeping the average age of Tynda's citizens at less than thirty. Despite the high natality rate, the BAM capital has very poor maternity and paediatrics services. Because decent food and whole milk are premium items, BAM children are sick for prolonged periods an average of three times per year, and 'there are some who are sick three times per month'.[48]

To attract volunteers to the BAM zone, the Soviet régime offered a number of material incentives including higher wages, longer vacations, priority rankings on automobile and apartment lists, and so on. All of these special benefits were contingent upon three years of contractual service. 'Some say a million migrated to BAM, each costing 20,000 rubles more than a common labourer in the centre. The cost of this migration was astronomical.'[49] Nevertheless, after all the expense, at the end of their third year of service, only one out of four labourers 're-enlists' for further tours of duty on BAM.[50] Together with their disaffection for the housing situation, 79 per cent are upset with the level of socio-cultural amenities, 53 per cent are dissatisfied with the child-rearing environment (little entertainment or learning facilities for children) and 50 per cent deplore the level of medical care. Yet, the 'most stable segment of the BAM population' is the worker with a family: among those who have lived on the BAM for more than six years 94 per cent are persons with families and only 3.4 per cent represents bachelors.[51] Thus, it would appear that BAM residents reside in a milieu that combines the worst of both environmental qualities: a poor human condition combined with a disrupted and deteriorating physical environment.

Is there a limit to the human ecumene? Has the Soviet government pushed it to its limit? True, human beings reside in even more hostile environments than BAM. During the 1980s, some 4 million people lived in the Soviet 'Far North' (north of 55° in the Soviet Far East):

> about 40 per cent live in mining districts, of which the most important are Kola, Vorkuta, and Noril'sk. Another 25 per cent are in coastal settlements based on military and fishing activities. About 20 per cent are concentrated along the major northern river valleys, and only 15 per cent – primarily aboriginal peoples – are scattered throughout the rest of the territory.[52]

The Soviet government has achieved this pattern of settlement at no small expense, and many, including Soviet economists, have questioned the wisdom of the massive subsidy required to create it. For example, elsewhere in the world (Canada, Alaska, Scandinavia and Greenland), the so-called 'Far North' has been developed in a limited manner because of the excessive expenditures inherent in northern development.

The BAM project differed from far northern regional development not only in terms of its geographic location but also in terms of the projected scale of that development.[53] The BAM railway was packaged with plans to engineer an entirely new region of the earth three times the size of France. For the first time in Soviet regional developmental history, a major project (BAM) was to be built comprehensively with environmental protection 'carefully' integrated into the plan. But, as James Bater has pointed out in his book *The Soviet Scene*, there is an enormous gap between Soviet ideals and realities.[54] BAM realities proffer a shocking array of negligence wth respect to both the physical and the human environment. Although BAM may still fulfil some of its promise, its environment and its people may not be able to wait. As this paper has revealed, for the time being life, if it can be called that, in the BAM service area is truly difficult to bear.

Notes

1. *Russkaya mysl'*, 2 November 1990.
2. Arthur Getis, Judith Getis, and Jerome Fellman, *Introduction to Geography* (Dubuque, Iowa: William C. Brown, 1988), p. 180.
3. George Frost Kennan, *Siberia and the Exile System* (Chicago: University of Chicago Press, 1958), pp. 22–6.
4. Donald W. Treadgold, *The Great Siberian Migration* (Princeton, NJ: Princeton University Press, 1957), p. 33.
5. Gary Hausladen, 'Russian Siberia: an integrative approach', *Soviet Geography*, 30, no. 3 (1989), pp. 231–46.
6. Treadgold, *The Great Siberian Migration*, p. 33.
7. James R. Gibson, *Feeding the Russian Fur Trade* (Madison, Wisconsin: University of Wisconsin Press, 1969), pp. 6–7.
8. Leslie Dienes, 'The development of Siberian regions: economic profiles, income flows, and strategies for growth', *Soviet Geography: Review and Translation*, 23, no. 4 (1982), p. 208.
9. 'BAM posle fanfara', Soviet television documentary (1988). Witnessed in Khabarovsk in May 1989.
10. *Gudok*, 18 September 1984.
11. *Gudok*, 6 January 1989.
12. *Gudok*, 4 September 1990.
13. *Gudok*, 6 January 1989.
14. *Gudok*, 17 July 1990.
15. A. I. Buzykin, M. A. Shary, and M. D. Yevdokimenko, 'Rubki glavnogo pol'zovaniya v tsentral'noi chasti zony BAM', *Lesnoye khozyaistvo*, 10 (1988), pp. 11–15; *Gudok*, 28 August 1988.
16. Robert G. Jensen, Theodore Shabad, and Arthur W. Wright, *Soviet Natural Resources in the World Economy* (Chicago: University of Chicago Press, 1983), p. 43.
17. *Gudok*, 13 October 1988.
18. *Gudok*, 18 October 1989.
19. *Gudok*, 12 December 1990.
20. *Pravda*, 25 July 1988.
21. *Gudok*, 18 April 1989.
22. *Gudok*, 7 March 1990.
23. *Gudok*, 13 October 1988.
24. *Gudok*, 7 March 1990; Vladimir Fedyakin, '*Golobyye goroda*: Kak slozhitsya sud'ba stroitelei BAMa i ikh semei?' *Ogonyok*, 5 (January 1989), pp. 14–17.
25. Yu. G. Melik-Stepanov and L. A. Chirkova, 'Tsvetnaya metallurgiya za desyatuyu pyatiletku', *Geografiya v shkole*, 5 (1978), pp. 8–9; A. A. Nedeshev, F. F. Bybin, and A. M. Kotel'nikov, *BAM i osvoyeniye zabaikalya* (Novosibirsk: Nauka, 1979), pp. 93–6; Paul Dibb, *Siberia and the Pacific* (New York: Praeger, 1972), pp. 132, 238.
26. *Gudok*, 27 December 1990.
27. Ibid.

28. V. V. Kryuchkov, *Chutkaya subarktika* (Moscow: Nauka, 1976); V. V. Kryuchkov, *Krainyi sever: problemy ratsional'nogo ispol'zovaniya prirodnykh resursov* (Moscow: Mysl', 1973); V. V. Kryuchkov, *Sever: Priroda i chelovek* (Moscow: Nauka, 1979).
29. *Gudok*, 13 October 1988.
30. Victor L. Mote, 'A visit to the Baikal–Amur mainline and the new Amur–Yakutsk rail project', *Soviet Geography*, 26, no. 9 (1985), pp. 691–716.
31. *Gudok*, 27 September 1989.
32. Fedyakin, 'Golobyye goroda', p. 16.
33. *Gudok*, 29 December 1990.
34. *Gudok*, 13 October 1988.
35. *Gudok*, 25 May 1990.
36. *Izvestiya*, 27 December 1986.
37. *Gudok*, 25 May 1990.
38. Mote, 'A visit', p. 710.
39. *Gudok*, 4 July 1990.
40. *Gudok*, 5 December 1990.
41. *Gudok*, 30 November 1990.
42. *Gudok*, 25 November, 2 December 1990.
43. *Gudok*, 9 September 1990.
44. Yu. T. Skorokhodov, 'Sovetskii Dal'nyi Vostok – problemy i perspektivy'. *Problemy Dal'nego Vostoka*, 2 (1988), p. 9.
45. *Gudok*, 4 September 1988.
46. *Sotsialisticheskaya industriya*, 11 October 1988.
47. *Gudok*, 9 September 1990.
48. Ibid.
49. Ibid.
50. *Gudok*, 22 November 1986.
51. *Gudok*, 9 September 1990.
52. US Central Intelligence Agency, *Polar Regions Atlas* (Washington, DC: CIA, 1978), p. 16.
53. Theodore Shabad and Victor L. Mote, *Gateway to Siberian Resources: The BAM* (New York and Washington: Wiley/Scripta, 1977).
54. James Bater, *The Soviet Scene: A Geographical Perspective* (London: Edward Arnold, 1989), pp. 1–2.

5 The massive degradation of ecosystems in the USSR

Zeev Wolfson

Global-scale destabilisation of ecosystems on Soviet territory

Six decades ago the Soviet empire was established on a huge territory larger than that of any other empire in human history. The Soviet Union possessed the richest natural resources in the world – fertile land, forests, fresh water, oil, metal, gold deposits etc. The idea of boundless space and inexhaustible resources was one of the main theses of Soviet ideology which created the illusion of the unlimited opportunities and superiority of Communism and, ultimately, of its triumph over all other social and political systems. The availability of huge natural wealth buttressed the false belief of inevitable success held by generations of Soviet people. On the other hand, the vast Russian territories, great rivers, lakes and seas where all pollution seemingly vanished, somewhat eased the USSR's ecological crisis in the last decades.

In the early eighties, however, this ideology was completely discredited, just as the resources of the huge Russian territories have been almost exhausted. In recent years environmental disasters have been increasing in the Soviet Union, so the question of the stability of its ecosystems and the forecasting of trends has become very topical not only for the Soviets, but for their European neighbours as well.

For an appraisal of the general ecological situation in any region or country, one should analyse the numerous social, economic, legal and technological factors. But one must not disregard the fact that the economic and social systems developing in a certain territory always have a geographical location. Each territory is a complex of numerous ecosystems whose specific features should be seriously taken into account. Experts deal with various aspects of fauna, flora, air and water. But from the point of view of environmental policy, the notion of stability of ecosystems plays a key role.

Unfortunately, research on ecosystem stability in the various Soviet regions is only just beginning, according to the Soviet geographer Preobrazhensky.[1] No books and few articles related to this subject have apparently appeared so far. Nevertheless, Soviet experts believe that stability can be defined as follows: 'Ecosystem X is considered stable as long as it is able to reproduce itself even though its parameters change within certain limits without impairing the natural habitat or damaging the health of the population'.

Soviet experts evaluate the stability of the ecosystems in the different regions of the country from a geographical viewpoint. I wish to refer to the 'Map of the acute ecological condition' which is the result of long research work by the Institute of Geography of the Academy of Sciences of the USSR.[2] Judging from the methodology of this research and the criteria established by the researchers, one can maintain that the term 'acute ecological condition' is virtually equivalent to the concept of drastically destabilised ecosystems or, in more general terms, ecologically disastrous areas.[3] On this map are demarcated 291 destabilised areas where the population encounters grave health problems related to various kinds of pollution. These areas constitute about 16–20 per cent of the territory of the USSR. Perhaps at first glance this figure does not seem so dangerous, but it relates to the entire territory of this huge country. For instance, in the European part which is the most densely populated, about half the territory belongs to the category of ecological disaster.[4] It includes deforested and former agricultural areas, vast tracts of erosion and soil pollution and regions of natural resources' extraction and industrial agglomerations. Besides the areas of acute ecological condition, the maps demarcate the areas affected by acid rain which also affects an additional 10–12 per cent of European Russia.

The Soviet experts give much the same priority to the dangers of soil degradation as a basic element of environmental stability as their Western colleagues. In order to obtain a more accurate estimate of the stability of the ecosystem one should note that the areas affected strongly by acid rain are adjacent to areas of ecological disaster and constitute a focus of their potential expansion. For instance, on the Kola Peninsula and around the city of Leningrad it is hard to find 'normal' ecosystems; almost all of them are classified as disaster areas or areas suffering from heavy acid rains. The Baltic republics and Belorussia are in a relatively better position, although this map does not include the radiation impact of the Chernobyl disaster. As admitted by Soviet and Western authors, its radiation affects millions of people in Belorussia and the Ukraine, and the contaminated area is spreading incessantly due to the propagation of the radioactive fallout by air, water and groundwater flows.

Radiation is a particular kind of environmental pollution, and we do not yet possess the criteria for evaluating ecosystem stability according

The massive degradation of eco-systems in the USSR   59

to the level of radiation. As is well known, the damage caused by radioactive substances can be observed only after a long period of time, while on the surface there is sometimes even an increase in biomass production. But from the point of view of human survival and the economic value of these regions, they should be included in the category of areas of environmental disaster.

Ecological degradation

The analysis of available Soviet data and studies allows us to take a broader view on the trends shown in the map of acute ecological situation. The most worrying trend is that the disastrous sites and 'islands' are spreading to the previously stable areas adjoining, thus forming a large territorial block. The significance of this process goes far beyond a mere arithmetical addition of the percentage of total territory in the USSR or any part of it. For this is a powerful ecological phenomenon extending the breadth of the disaster areas. For instance, instead of the 15–20 scattered areas of acute environmental pollution in the Upper and Lower Volga in recent years, practically the entire Volga basin has become an ecological disaster zone.

The situation in the centre of European Russia cannot be isolated from the developments in the other threatened areas – the Far North on the one hand and the South on the other. But let us consider the mechanism of expansion of ecological disaster before we look at these geographical regions in more detail.

The most interesting question is the question of stability of the adjoining territories, which are still productive enough. Under the heavy impact of the disaster zones, however, they have somehow become unstable. This impact can spread in the following ways:
(a) fall of the groundwater level and water contamination;
(b) strong erosion or salination of the soil;
(c) decrease of biological productivity (due to the effect of both the above-mentioned factors);
(d) depletion of various species of flora and fauna with a substitution by more primitive ones more adaptable to severe conditions, for more sophisticated, and more developed ecosystems;
(e) changes in the physical condition of the environment.

If an ecosystem is affected to a significant degree by some or all of these factors, its stability decreases and a few additional destabilising endogenous or exogenous factors – like drought or overgrazing by cattle – are enough to cause the productive ecosystem to collapse. What is more, the impact on the neighbouring, still 'normal' system somehow increases and the process of degradation continues. Sometimes it may be unnoticed, but another year of drought or an ill-conceived project of irrigation or industrial development provides one more 'push' – and another ecosystem collapses. The new ecosystems differ from those already in existence not only by their lower productivity and more primitive structure but by their reduction of economic and social potential. Since they are unable to regenerate, this makes the reproduction of a normal human population (which has lived there for centuries) and the continuation of their ethnic and cultural traditions doubtful.

The Sahara and the areas situated to its south are a case in point. Roughly 1.5 million hectares of land turn into desert there annually while each year up to 6 million hectares deteriorate worldwide almost to the point of non-recovery. The trends in Soviet Central Asia in the last few years are less known, but the absolute figures for desertification and environmental degradation in North Africa and Central Asia are similar, and it is known that ecological disaster zones in Central Asia are spreading more rapidly than those in the Sahara-Sahel.

Although specialists study various kinds of pollution in their different separate aspects – water, air, soil, flora and fauna – various ministries may exercise their control over each particular aspect (which is actually what they do in most cases). In practice, however, desertification is the end result of mass pollution and contamination of the environment. Desertification constitutes the combined effect of all visible and invisible disruptions of natural resources.

The Aral ecological catastrophe is well known to Western scholars. But it would be misleading to consider that the drying up of this huge natural body, the resulting health hazard and loss of employment are the ultimate consequences of this catastrophe. Now with 42 per cent of the bed of the Aral Sea dried up, strong winds lift up approximately 200,000 tons of sand and salt *each day*. The heavier sand particles settle within a radius of 400–500 kilometres, while the lighter salt particles are carried to much greater distances. Satellite photographs show salt storms extending up to 10,000 kilometres to the south and west. They reach the northern Caucasus and Belorussia and form layers of dust and salt over the glaciation of the Pamirs and Tyan' Shan' which feed the two largest rivers of Central Asia. This speeds up the melting of the ice, raising somewhat the water level of the Syr-Dar'ya and the Amu-Dar'ya temporarily, but only temporarily. When the many-metres thick, millennia-old ice mass has melted, the irreplaceable mechanism regulating the river flow will disappear, probably for good. And the overall water supply will decrease sharply, especially during the critical summer season.

The senseless waste of water did more than dry up the Aral Sea; it also caused the salination and desertification of enormous areas of land. 'If you drive along the Khiva-Bashauz-Nukus route', wrote Corresponding Member of the Academy of Sciences of the USSR, A. Monin, 'you will see snow-white steppes, a lifeless plain, marshes stretching as far as the eye can see. We must face the truth: it looks as if these salt marshes are here for good.'[5] According to A. Reteyum, a prominent Soviet geographer, throughout most of Central Asia the biological productivity of the ecological systems has declined by 30 to 50 per cent, many rivers and lakes have disappeared altogether, and the total area of sand deserts has grown by over 100,000 square kilometres in the last few years. Reteyum sees all the signs of deterioration as being beyond human control.[6] Indeed, the Kara-kum and the Kyzyl-kum, the ancient 'nuclei' of Central Asian deserts and semi-deserts, occupy roughly 700,000 square kilometres. This is about 22–24 per cent of the total area of Uzbekistan, Turkmenia, Kazakhstan and the adjoining territory to the north of the Caspian Sea. But now active desertification has raised the total to approximately 35–40 per cent of the area. Each year the desert swallows up from 800,000 to 1,100,000 hectares in the very heartland of the Asian continent.

Another Soviet expert, S. Zabelin, calls the processes developing in the Central Asian region 'Sahelization'.[7]

Further degradation

Of course, the extension of desert in Central Asia takes place in other directions, not only to the west, but in this chapter we are focusing on the Europe–Asian relationship. Therefore, particular attention should be paid to the Tengiz area, the Volga delta and the Kalmyk steppes. These three regions are areas of expanding environmental disaster. At the same time they constitute a kind of bridge which links Central Asia with the southern Ukraine, a European territory in the full geographical and psychological sense of the word.

Tengiz is a region along the Caspian Sea in western Kazakhstan (between Gur'ev and Shevchenko) where Soviet enterprises and American firms are beginning to exploit oil and gas deposits. As a result of the vulnerable semi-desert environment (which implies great instability of the ecosystem) and the difficult conditions of oil extraction, grave degradation of the environment is already spreading across thousands of square kilometres even before the main industrial projects have begun.

Owing to Russia's acute need for new sources of energy, the probability is high that despite the efforts of the Soviet Greens and the ecological lobby, a new vast area of man-made desert will arise in the next three to five years. Less than 300 kilometres to the north-west of the Tengiz region is the great Volga delta, three sections of which form a nature reserve. The development of the chemical and gas industry, together with the heavy pollution of the Volga River, has turned the Volga region in the last few years into an ecological disaster area which has seen a sharp increase in diseases and the growth of mutations among animals and man.

In Kalmykia, as monitored by Soviet experts, reckless year-round pasturing of cattle, formerly done on a seasonal (winter) basis, has led to the formation of the first sand desert in Europe. Its present size is estimated at 7,000 square kilometres and increasing annually by 10 per cent. Despite warnings from specialists, the misuse goes on and, as Soviet geographers write, the Kalmyk steppes 'are systematically being transformed by man into a sand desert'.[8] If the process of desertification maintains its pace, it might affect regions as far to the west as the southern Ukraine within the next five years – all the more likely since these regions situated between the Kalmyk steppes and the River Dnieper are already overloaded with industrial and agricultural pollution. Yet if one frontline of eco-degradation moves westwards in the south of the Soviet Union, another one moves to the north.

According to the classification of the Soviet geographers Yu. Golubchikov, Yu. Solomatin, etc.,[9] the ecosystem of the tundra, forest-tundra and coniferous taiga is even less stable than that of the semi-deserts. They base their conclusions on the high vulnerability of the northern ecosystems, referring both to the historical process of formation of these systems and to the present enormous economic impact on these regions.

In addition to these very serious transformations in the Far North, there is a serious threat from the advance of the sea towards the Arctic coastal areas. For instance, in the Yamal Peninsula, where the altitude above sea level is only a few metres and the development of the gas industry constitutes a threat to the peninsula of total destruction by floods. The peninsula with its total area of over 100 thousand square kilometres could simply disappear beneath the sea . . .[10]

The powerful impact of the collapsed northern ecosystems on the region situated to south and west is quite significant due to the scale of the tundra and coniferous taiga zones. Within the geographical borders of the European USSR these two zones occupy over one million square kilometres or more than 15 per cent of the Soviet territory west of the Urals.

Moreover, between the Urals and Central Russia there are no mountain chains or other areas which can act as a natural barrier slowing down the degradation process.

Conclusions

Firstly, one must point out the dangerous processes of the massive degradation of ecosystems in the south of Russia as well as the Far North and the even worse deterioration in Central Asia.

The situation in so-called Central Russia is also described as a disaster in numerous areas. Thus there is good reason to speak of the formation of an *united front of ecological degradation between Scandinavia and the Black Sea*.

Secondly, the formation of huge territories of ecological deterioration has had far-reaching consequences. The process of degradation has its own dynamic in each given region, but the most important point in my analysis is that the total of two or more processes in the merging disaster areas may be worse than the sum of its parts. Of course, their effect will be felt above all in the disaster zones and the neighbouring territory of European Russia. But later they will seriously accelerate environmental degradation in Eastern and Central Europe.

Within the framework of my approach, I give first priority to qualitative analysis in order to draw attention to the problem itself. This may help in the further qualitative evaluation of the extent of massive ecological destabilisation in the various parts of Europe.

Notes

1. Quoted from Yu. Golubchikov, 'Ustoichivost Severnykh Ekosistem k Antropogenomu vozdeistviyu', Seriya *Znanie RSFSR* (Moscow: 1990), p. 5.
2. B. Kochurov *et al.*, 'Osnovnoye soderzhaniye karty ostrykh ekologicheskikh situatsii', *Prirodno-Ekologicheskiye Sistemy* (Moscow Division of the Geographical Society, 1989), pp. 30–41.
3. Ibid., p. 36.
4. *Izvestiya*, 26 March 1990, p. 3.
5. A. Monin, 'Zastoinye zony', *Novyi Mir*, no. 7, 1989, p. 165.
6. 'Srednaya Aziya i Kazakhstan', *Kommunist*, no. 14, 1989, p. 33 (speech by A. Reteyum).
7. S. Zabelin. Personal communication, March 1990.
8. V. Kotlakov *et al.*, 'Sovremennoye Sostoaniye i Okhrana Prirodnykh Resursov Yugo-Vostoka ETS', *Vestnik AN SSSR*, no. 5, 1988, p. 62.
9. Yu. Golubchikov, Yu. Solomatin, 'Metodicheskiye aspekty ustoichivosti Severnykh Ekosistem', *Issledovaniye Ustoichivosti Geosistem Severa* (MGU, 1988), pp. 20–6.
10. Ibid., pp. 5–6.

6 The new politics in the USSR: the case of the environment

Joan DeBardeleben

In this chapter, I attempt to unravel the changing characteristics of politics under perestroika by examining the environmental issue in greater depth. My goal is to produce some general propositions about the changing nature of politics up until late 1990 in the USSR. The area under consideration, environmental politics, is appropriate as a kind of case study for several reasons.

First, in the pre-perestroika era, this was an issue characterised by *relatively* free discussion, but still within important constraints characteristic of the period.[1] Because discussion was relatively free, the nature of political conflict was also *relatively* more transparent than for some other issue areas (e.g., nationality politics, foreign policy).[2] The greater openness of conflict and debate in this area probably stemmed from the perception that environmental proponents were not fundamentally challenging the foundations of the prevailing ideology or power structures. In cases where they did, for example in questioning the desirability of economic growth as a systemic goal, or in challenging the safety of the state's nuclear power strategy, restrictions on debate were enforced.[3] Likewise, information about the disastrous state of the Soviet environment was suppressed.[4] None the less, within fairly broad parameters the scope of discussion was significant, and its range has expanded rapidly since the mid-1960s.

Most actors in the political conflict were élites, if this term is broadly taken to include upper ranks of the intelligentsia. This included, on the pro-environmental side, representatives of the intelligentsia (especially writers, economists, journalists) and of state environmental regulatory organs. The greatest resistance to change came from industrial interests and the party–state economic apparatus.[5] While some mass-based conservation organisations existed (for example, the All-Russian Society for the Conservation of Nature), these groups were not involved in mass political mobilisation of the population. The environmental area was, however, unique and in this sense a harbinger of more recent patterns, in that at least some environmental initiatives came from society

The new politics in the USSR 65

(expressed by writers or journalists) rather than from the party–state hierarchy.

In the Brezhnev period, a large amount of legislation was passed on environmental matters, as on many other issue areas. However, the implementation of these measures was largely ineffective. Apart from bureaucratic resistance from the more powerful productive ministries, the old economic structures and mechanisms presented major obstacles to an effective environmental policy. For example, the priority of the issue was low, particularly in light of the traditional Soviet emphasis on sectors generating material output. Incentives reinforced this emphasis on meeting output quota. Free use of natural resources, rooted in an interpretation of the Marxist labour theory of value, inhibited their conservative use.[6] And administrative structures failed to separate functions of use and protection. Thus, despite a wide body of legal enactments, environmental policy was largely ineffective in halting resource waste and pollution.

These features of environmental politics before Gorbachev suggest that while the bases of political conflict were more transparent than for other issue areas, it was typical of many domestic issue areas in several regards: (a) the inter-linking of political and economic concerns (e.g., the rooting of policy inefficacy in the centralised economic structures and the key role of industrial ministries and large enterprises in undermining effective implementation); (b) the flexibility of the system in allowing policy debate which did not undermine basic ideological principles, systemic stability, or major economic priorities; (c) the minimal role of the mass public as political actors; and (d) the generally centralised nature of decision-making.

In the Gorbachev period, especially since 1986, the nature of politics in the environmental area changed significantly. Most importantly, the issue became highly politicised, on a mass level.

On the surface, two factors explain this development. First, the Chernobyl nuclear accident in April 1986 drew attention to the very real environmental dangers in the productive system and made evident the unreliability of official reassurances about environmental safety.[7] Following the initial burst of information about the accident (the first real evidence of 'ecological glasnost'), disclosures of public misinformation have continued until late 1990. (Restrictions on the press being implemented at the time of writing, in early 1991, suggest that this may change.) The Chernobyl accident spurred widespread opposition to the further construction of nuclear power plants and this activism in several cases resulted in the closure of plants or construction stops on new ones.[8]

A second proximate stimulus to environmental activism was ecological glasnost. While public information was still considered inadequate by many experts in the late 1980s,[9] it became much more extensive than previously. At least through 1990 scientists were able to publish data in the mass media which was previously considered secret; this in turn helped to sensitise the population to the hazards around them. While citizens may previously have suspected high levels of pollution and consequent health effects, official reassurances muted any response. The availability of published data from established scientists legitimised and activated latent fears and suspicions.

While these two factors, the Chernobyl accident and ecological glasnost, were the primary immediate sources of the rising salience of environmental issues, I will be arguing that the deeper causes are symptomatic of the changing nature of politics in general in the USSR.

What has been the policy response to this growing environmental activism? As in the Brezhnev years, several legal enactments have been put into effect since 1985. The most important of these are the January 1988 resolution of the CPSU Central Committee and the Council of Ministers on restructuring state environmental organs (creating Goskompriroda, the new State Committee for Environmental Protection), and a November 1989 resolution passed by the Supreme Soviet. As I have treated these and other policy innovations elsewhere,[10] I only summarise the general patterns here.

First, as in the past, the numerous measures adopted have been ineffectively implemented. Changes in the economic incentive structure for enterprises, envisaged in the Law on the State Enterprise adopted in 1987, were not realised, in part because the economic reform as a whole faltered. In addition, the environmental component has not been of markedly high priority within the economic reform. Likewise, although a new environmental agency, Goskompriroda, was created, it was confronted with numerous problems – ranging from inadequate staffing to an unclear range of authority. Its efforts have not been entirely ineffectual, but less effective than envisaged. Second, many elements of the reform package which might have a positive environmental impact did not yet come to fruition, most notably a strong decentralisation of economic decision-making to the level of the local soviets or the Union Republics. While local elections put in place some new local leaders with a clearer mandate to address environmental issues, their real financial and legal resources and powers remain limited. Until now local soviets and local citizens' groups have had to appeal to higher state authorities to realise environmental goals.[11] While in many cases these appeals have met with a favourable response, the powerlessness of local organs has been a source of frustration for environmentalists and has inhibited a real process of efficacious decision-making at the base level of the political system.

Policy response has been visible mainly in the numerous *ad hoc* responses to popular pressure focusing on particular environment 'crises'. This pattern resembles the Baikal syndrome, where vocal environmental advocates at the local level were able to elicit a policy response from the centre. From 1987 to 1990, all across the USSR, plants were closed, planned projects were re-sited or re-tooled for a less polluting type of production, or projects were cancelled altogether. The most prominent examples include the cessation of work on the planned river diversion projects, cancellation of the Volga–Chograi canal, closing of biochemical plants, and plans to convert the Baikalsk Pulp and Paper Plant to furniture production.[12] Numerous smaller victories also occurred elsewhere. These decisions in some cases satisfied local activists. However, they did not involve a thorough public evaluation of the costs of alternative responses. In the fall of 1990, Gorbachev ordered the reopening of some of the polluting facilities due to the unacceptably high costs incurred by their closure.

In sum, the policy response has brought some particular 'victories' for environmentalists, but has not brought any structural change which might ensure a longer range improvement in the environmental situation in the USSR. On the other hand, by 1990 the pro-environmental coalition was stronger than previously and this suggests possibilities for future structural change.

Coalition patterns: old and new actors

Previous Western research allows us to develop a schematic representation of environmental coalitions in the Brezhnev era.[13] On the left-hand side of figure 5.1 are listed those forces which generally took positions in support of environmental protection, while on the right are indicated those which placed priority on economic development over environmental goals. In the centre column are groups which were cross-pressured and were generally ineffectual sources of support for environmental causes. While this scheme is in some ways overly simplistic (since, for example, scientists and experts may disagree over the environmental impact of contentious projects), it provides an adequate point of departure for our discussion of environmental coalitions under perestroika. It suggests that pro-development forces on the whole were considerably stronger than the pro-environmental forces in the Brezhnev period.

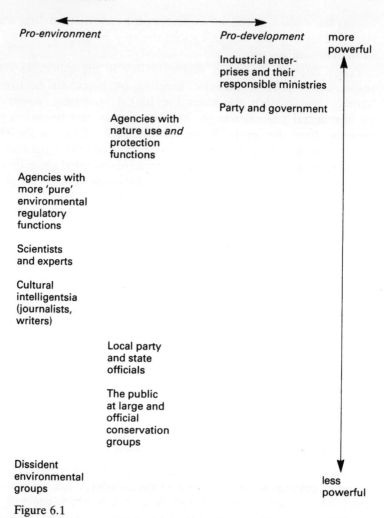

Figure 6.1

The cross-pressured groups require some special attention. For example, a conflict of interest was inherent in the mandate of many agencies charged with nature-protecting functions during the Brezhnev period, for these organs often had responsibility for both the protection and the use of nature. For example, the Ministry of Land Reclamation and Water Resources had some responsibility for controlling water and soil pollution. In practice, however, it sponsored major water construction projects (including the much-discussed river diversion projects) which environmentalists consider to be serious sources of ecological disruption.¹⁴ Likewise, local organs faced conflicts of interest. To bring improvements in the local environment, they were heavily dependent on productive ministries for resources. Furthermore, their performance was often judged primarily by the success of the productive enterprises in their region. As an example, it was previously considered a feather in a local leader's cap to have a nuclear power plant constructed in the region. If local leaders sought to challenge polluting enterprises in their locality, they had to seek recourse to higher party organs or the responsible ministry. Finally, the semi-official conservation societies were not independent of state control and thus played only a minimal role in the policy-making process. They were important primarily for educational work done at the local level.¹⁵

While the environmental lobby did exist in the Brezhnev period, it was weak due to the primacy of production goals and institutions in the system and dependence on the organs responsible for material output. Pro-environment forces had few resources except scientific expertise. But censorship prevented environmental advocates from using information to mobilise popular support for an environmental agenda.

With the advent of glasnost, this situation changed dramatically. New actors appeared and organisations which were previously cross-pressured were affected by new factors. Enterprise incentives also changed and, with the addition of cooperatives to the productive and service sectors, the productive enterprises became a less unified actor. The three most important new actors have added increased weight to the pro-environment coalition. These are: (a) the media, which previously was largely an instrument of other actors, but under perestroika took on an independent role in the political process; (b) unofficial public organisations formed from below and answerable to their adherents; and (c) the new legislative bodies, in the first instance the Congress of People's Deputies and the USSR Supreme Soviet, but then joined by similar organs at republic and local levels. Unlike their counterparts in the Brezhnev era, the new soviets were elected on a competitive basis, have the power to challenge other state bodies (including the ministerial apparatus and state enterprises), and are, to a significant extent, accountable to the public and not to the Communist Party. In addition, important powers in the environmental area have been claimed by republic level organs.

In addition to these new actors, the stance of those forces in previously ambiguous positions is undergoing change.

Environmental regulatory agencies

The restructuring of environmental agencies has produced a less fragmented regulatory structure and, with the creation of Goskompriroda, has helped to separate functions of nature use and protection. Thus, Goskompriroda, which has no productive functions, falls clearly on the side of environmental advocacy. While this organ is still of relatively low efficacy, it, and its republic-level branches, have been instrumental in closing some highly polluting facilities and in raising public awareness of environmental problems. There is no evidence that it has become a captive to the mentality or influence of those whom it seeks to regulate. This may of course happen to some degree in the future, as has occurred in some Western polities. Some citizen activists quite clearly look to Goskompriroda and its regional organs as allies;[16] others are critical of Goskompriroda and express concern that bureaucratisation may make it increasingly ineffectual.[17]

Goskompriroda has close links with both state officials and social forces.[18] The committee itself has about fifty members, including ministers and deputy ministers; it meets two to three times a year to discuss environmental protection policy. The committee is chaired by Nikolai Vorontsov, whose credentials are scientific, not administrative; he is not a Communist Party member. Attached to the committee, in an advisory capacity, are the Societal Council and the Scientific-Technical Council. The Societal Council has 100 to 150 members drawn from a wide range of backgrounds, including representatives of religious organisations, the scientific community and youth. Its functions are to provide a link to public opinion and to make recommendations to the Committee itself. It meets publicly about twice a year and is represented in between by its bureau, which has about thirty members. By chance, about six members of the Congress of People's Deputies are also on the Societal Council. (They are not *ex officio* members.) This council is also divided into working commissions dealing with specific questions such as international relations and nature reserves. The Scientific-Technical Council has about thirty members and also serves the State Committee in an advisory capacity.

Apart from the two councils, the State Committee has a board which heads five divisions to deal with the following areas: economics and the organisation of nature use; science and ecological normatives; monitoring; the management of the propagation of ecological knowledge; and ecological *ekspertiz* (assessment of environmental impacts).

As this brief description indicates, Goskompriroda has a large area of responsibility, distinct from any productive functions; it also has a mandate to maintain contact with larger social forces involved in environmental affairs. Ironically, however, just as nature protection functions became more concentrated (with the formation of Goskompriroda), pressures for republican sovereignty increased regional fragmentation in establishing and implementing environmental policy.

Local state and party organs

The position of local organs (both state and party) remains ambiguous for two reasons. First, until mid-1990 and in most cases after that, these organs were still staffed largely by old personnel, selected under the influence of the nomenklatura system. Only with the competitive local elections of early 1990 did local officials and deputies become accountable, in any sense, to the public. Despite new legislation on local self-management and the local economy, local organs have only minimal economic authority and fiscal independence, a *sine qua non* for effective local action in the environmental area.

Even with increased fiscal authority and public accountability, the ambiguous position of local authorities might not disappear, however. For local authorities must concern themselves with the quality of life more broadly. Environmental protection is only one component of this. Whether local enterprises are subject to central directives, *khozraschet* (self-accounting), or the market, they must balance the achievement of productive and ecological goals. While economic development and ecological goals are not mutually exclusive, in the shorter term there may be real trade-offs between increases in material affluence and maintenance of ecological balance. Furthermore, some of the most troubling environmental issues, such as the development of adequate sewage disposal systems, represent a drain on strained local resources. The support by Central Asian party and state officials for the Siberian river diversion projects is an excellent example of such conflicting pressures. Water from the proposed diversions would help ameliorate local water shortages and thus bring potential improvement in local conditions, even if broader ecological effects might be negative. Thus, local authorities are unlikely to become unambiguous advocates of environmental protection. Glasnost and perestroika made the contradictory position of local organs more transparent, as local authorities face conflicting public pressures.

In the last two to three years, local authorities have at times supported environmental activists,[19] but in other cases they have placed obstacles in their path. Local officials often seem to see their interests as closely linked to the traditional developmental priorities of the ministries represented in their regions. For example, citizen activists who organised to demand the cessation of harmful emissions from the biochemical feed protein factory in Kirishi (near Leningrad) confronted a united coalition of the Ministry of the Biomedical Industry and the city soviet executive committee.[20] Likewise, when citizens in the Brateyevo region of Moscow organised their own self-government committee to protest against the building of several new enterprises in their already polluted district, the Krasnogvardeiskii District Soviet and Soviet Executive Committee initially failed to support the movement; only after the local activists organised an independent referendum of residents to choose between their developmental plan and the city's did the district soviet support the outcome (in favour of the activists' alternative plan) before the Moscow city soviet.[21] Other cases where local environmental activists faced obstruction from city officials include efforts of a Ryazan citizens' group to preserve a local natural monument and to protect the Oka River from pollution which they felt would result from the city's plans to route a sewage collector system through the area in question.[22] Likewise, in Ufa, citizens protested secrecy on the part of local officials regarding environmental and health effects of local pollution.[23] The ambiguous position of local officials is evident in the conflict over the construction of the controversial hydro-electric station in the Altai, for the authorities saw money for new housing and social facilities as dependent on the project, even if it might be harmful to the ecosystem and the health of the population.[24]

Increased financial independence would reduce the dependence of local authorities on productive ministries, but they still would face the multiple and varying interests of their constituents.

The media

The media have become independent actors in the environmental battle, but they do not always support environmental causes. Local newspapers have often continued to support the status quo, reflecting the viewpoint of the local and party–state apparatus.[25] None the less, alternative media sources (including central newspapers and magazines) have often provided outlets for environmental advocates. Several publications, including *Novyi mir*, *Yunost'* and *Sovetskaya Rossiya*, have sponsored particular environmental initiatives and expeditions.[26] The media also act as a clearing house for information on particular topics. In this context, one should not forget the electronic media. Here shows like 'Vzglyad', the weekly documentary programme, presented graphic evidence of environmental deterioration until the show was cancelled in early 1991.

The new soviets

Environmental issues played a prominent role in the election platforms of many candidates for the new Congress of People's Deputies in the early months of 1989,[27] as well as in the local elections in 1990. The issue was mentioned frequently in the early meetings of the congress as well as at the special 19th Party Conference held in 1983.

The new central legislative bodies have also provided an opportunity for environmentalists from all over the country to interact. An ecological caucus was formed and was able to influence the process of approval of appointees to head ministries and state bodies. The new state environmental protection agency, Goskompriroda, is now in a meaningful sense accountable to a truly public body, even if the process of legislative oversight is not well developed (nor well staffed) in the USSR. The Supreme Soviet's Committee for Environmental Protection and the Rational Utilisation of Natural Resources deals with environmental issues in a specialised manner. The committee includes prominent environmentalists and scientists, such as Aleksei Yablokov.[28]

Thus the new soviets offer a potential forum for discussion of the numerous choices which must be made in determining the priorities and the approach to environmental issues. This potential has not yet been realised, since the organs have so many pressing problems to confront and are comprised of deputies with virtually no expertise or experience in governing.

Unofficial environmental organisations

The leading specialist for social relations at Goskompriroda has estimated that in the spring of 1990 over 300 environmental groups existed throughout the USSR.[29] There are probably actually many more. Most are organised on the local level, often focusing on particular environmental 'hot spots'. They generally have a core of two to five individuals and a broader membership circle of twenty to thirty activists. They can often rally several hundred citizens in support of particular initiatives (rallies, demonstrations, petitions). Because new groups are constantly forming and disappearing, it is difficult to maintain an accurate overview of their size and numbers. Other organisations are now forming at the republic level, especially in the European republics. Environmental organisations also exist in Central Asia, but they tend to be weaker, despite the numerous ecological problems facing the region.

At the All-Union level, linkages between groups are only beginning to form. This is partly due to the fact that the movement has just recently erupted spontaneously from below and thus has not had time to coalesce (despite the formation of some umbrella groups such as the Social-Ecological Union). The movement is also fragmented for other reasons. First, in many cases the ecological issue is linked to other causes, such as ethnic nationalism, cultural revivalism, or even religion. Thus, for example, while the environmental issue was an important component of the platforms of the Baltic People's Fronts and associated organisations, positions on independence for these regions have taken on greater salience and may make difficult a coalition among environmentalists holding different positions on the national question. Second, the environmental movement, as in the West, is characterised by ideological splits. Groups such as Pamyat' link environmentalism to Russian cultural revivalism in a form repulsive to other wings of the environmental movement. While it is relatively easy to bring people with diverse ideological assumptions together in opposition to a particular polluting factory, it becomes more difficult to develop a more general policy platform uniting such diverse ideological trends. Western environmental movements have also been plagued by these strains, and there is no indication that the Soviet Union will escape. These strains may in fact prove to be more difficult due to the importance of ethnic tensions and the general lack of ideological consensus following the decline of Marxism-Leninism as a predominant ideological framework.

On a socio-economic level, it appears that the movement is largely led by members of the cultural and scientific intelligentsia, but draws its support from a broad socio-economic and age range.[30] Of considerable interest and importance is the position of industrial workers. In some cases, workers have joined themselves to the environmental cause in a relatively organised manner;[31] in other cases the implied threat to jobs from plant closures has led to public opposition to the demands of unofficial environmental groups.[32] If market reforms increase the threat of unemployment, contradictory pressures on the industrial workforce (between their role as employees and consumers of the environment) are likely to increase.

Without doubt, the new public organisations have been the major force behind the long list of 'ecological' plant closures and project cancellations. These new groups often turn to scientists and experts to provide support and substantiation for their claims; sympathetic newspapers and journals offer crucial publicity. The role of previously-existing official conservation societies has also been surprisingly visible,

The new politics in the USSR

as some of the new citizens' groups have formed affiliates of these organisations.[33] On the other hand, there have also been cases where the new activists have resisted joining forces with the established public organisations for fear of being coopted into a bureaucratic game.[34]

Enterprises

Even now some enterprises have joined an independent organisation, the Environmental Business Association.[35] This independent organisation has a formal contract relationship with Goskompriroda; it tries to further the production of ecological equipment, particularly through contact with foreign firms. In mid-1990 members in the organisation included over forty large enterprises as well as some academic institutions (e.g., Kazan State University, the Moscow Chemical-Technical Institute). Member enterprises seemed motivated by a desire to further contacts with foreign partners in the development of ecologically sound use of technology. Some forward-thinking managers were responding to growing public pressure for environmental responsibility.

By and large, however, state enterprises continue to resist demands by environmentalists. Old personnel and management mentalities survive. More importantly, the incentive structure has thus far not fundamentally changed. Self-accounting and/or market reforms may indeed increase pressures on managers to externalise environmental costs in order to improve profits. On the other hand, there are firms in the cooperative sector which are filling gaps in the production and repair of environmental equipment.[36] For example, a group of cooperatives in the food sector has underwritten an investigative expedition by *Yunost'* magazine to study environmental crisis zones in the USSR. *Yunost'* sees this type of sponsorship as a means of achieving financial support for its initiatives without being beholden to the state ministerial and enterprise structure.[37]

If marketisation proceeds, one can expect the role of the business sector (particularly the cooperative-private sector) to be increasingly complex. Pressure to ecologise production my well increase, as popular awareness grows; enterprises which reduce waste (and thus pollution) may also be more competitive. On the other hand, as noted above, externalisation of costs may also be increasingly tempting.

A broader view: environmental politics and Soviet politics

While several factors are increasing the weight of the environmental lobby in the USSR, in a broader view prospects for the Soviet environment are still not good. The environment faces tough competitors for public attention, money, and energy (for example, the economic crisis with its related shortages of almost all goods, and the nationality conflict). But the problem also lies in the evolving nature of Soviet politics itself. In the pages which follow, I will explore five dimensions of the new environmental politics in the USSR, with the purpose of illuminating some more general patterns in the transformation of Soviet politics. While these points are somewhat speculative, I hope they provide a starting point for conceptualising the dynamics of Soviet political conflict in this transition era.

The politics of passion

One might ask, why have environmental concerns been so much more susceptible to political action than some other commonly shared demands, such as, for example, consumer interests? Apart from the national issue, few, if any other issue areas have generated more vocal grassroots organisation than the environment. While ecological degradation itself certainly has an important impact on the quality of life of the average citizen, several other issues have a more or equally visible impact and yet have not produced the same level of public activism. Earlier I identified some of the more proximate causes of environmental activism in the USSR (Chernobyl, ecological glasnost); here I would suggest that the salience of this area reveals some fundamental characteristics of the changing nature of politics in the USSR.

At first glance, explanations for the green phenomenon in the Western context do not seem to apply to the Soviet case. Much of this Western literature emphasises a shift to non-materialistic, post-industrial values or a change in the nature of the productive structure toward a more service-oriented economy.[38] If anything, the turn toward 'green' politics in the USSR is occurring when the material standard of living is falling and where the term 'post-industrial' society is less applicable than previously. Material values are still very important in the USSR and affluence has, if anything, declined.

On a deeper level, however, there is a parallel with the Western experience. For the attention to environmental issues in the USSR reflects a revival of cultural and human values, as opposed to past policy which was overwhelmingly oriented to the production of material output. The citizen confronts visible damage in the immediate environment, and this damage is seen as a result of centrally-mandated production priorities; these priorities involved an overriding concern with producing material output (rather than services or human values). I am not suggesting that Soviet authorities have been indifferent to human values; this certainly was not the intent of the initial ideology nor probably of the actual leadership. But the incentive structure in the economy has rewarded material output, so much so that 'non-productive labour' (i.e., that which does not produce material output) traditionally has not even been accounted in measures of economic growth. Nor has natural wealth been given a concrete 'value', and natural resources themselves have been considered 'valueless' in an economic sense. The issue is then symbolic: desecration of the environment is perceived as desecration of one's culture, of one's humanness and innate worth. This connection is very clear in the platforms of the popular fronts of the national minorities, but also is evident in Russian areas where preservation of national monuments is linked to preservation of the environment. For national minority groups the destructive effects of activities by centrally-dominated industrial ministries represent a destruction of basic cultural values (including the value of the native land and nature);[39] in the Russian Republic, for instance, the centralised ministries are blamed for elevating narrow-minded, materialistic departmental interests above fundamental cultural values of the population.[40] The fact that the materialism of the productive structure has not even succeeded in producing adequate supplies of goods for the population makes the destructive effects of the productive system seem even more unjustifiable. In sum, the new 'green' phenomenon in the USSR is rooted in a rejection of the misguided materialism of the old structures. Like post-materialism in the West, Soviet environmentalism represents the ascendance of non-material, cultural, human values. Likewise, the language of the environmental movement contrasts sharply to the official class-based language which justified previous state economic policy. In some ways then, it is perhaps easier for the population to embrace the language of environmental politics (demands for a livable, healthy, natural environment) than to embrace oppositional class politics.

A second characteristic of the environmental issue may help to explain its salience. While environmental problems clearly have an important global dimension, they are in the first instance localised, visible problems in the immediate surroundings. They affect the smell in the air, the fish in the river and the health of the children. Even more importantly, the source of the immediate problem is visible and local – it is a specific nuclear power plant, a particular smokestack or an emission source into the river. Thus opposition to polluters is susceptible to relatively easy organisation and mobilisation on the local level. The target is clear, and a victory is both conceivable and achievable (closing of the plant). And indeed victories have been achieved in other localities; each round of victories increases the sense of political efficacy in a system where efficacy is scarce. Compared to other troubling issues (for example, the poor quantity and quality of consumer goods, systemic obstacles to upward mobility for women, the lack of adequate housing), minimal mobilisation may bring a visible improvement.

In sum, the environmental issue provides the possibility of joining passion with action. The issue is symbolic of a larger set of concerns relating to the cultural and human integrity of the population; it is possible to identify specific targets susceptible to local pressure by informed grass-roots movements. While this particular combination may be atypical of other issue areas in the USSR, the attraction to particularistic causes seems to be more universal. The ethnicisation of politics is another example, for here again a revival of particular cultural, linguistic identifications is involved. The old universalistic ideology of class politics, propagated by the party–state élite, has been largely de-legitimised (in part by the reform leadership's own reform programme), and yet there has not emerged an alternative ideology to replace it. There has been no legitimate redefinition of socialist values which can now form the basis of an alternative, universalistic ideology. In this context the appeal of particularistic interests is rising. This is not to say that class politics or the language of social class is dead in the USSR – they may well revive in some form, stimulated either by a resurgent party apparatus seeking popular support, or alternatively, by radical economic reforms and the concomitant social side-effects (unemployment, inflation, job insecurity, increased demands for job security, greater pressure for job discipline at the workplace). Already there are manifestations of class politics, especially in the form of miners' strikes. Until now, however, many of the demands of workers have related to broader political or quality of life issues, including the environment, rather than to workplace concerns. In the longer term, class-based issues may well revive, but in the transition period, the politics of passion, based on basic definitions of cultural and human identity, have taken precedence. The environmental cause is a paradigmatic case.

Symbolic politics: revised edition

Public rhetoric has had largely symbolic significance in the USSR in the last several decades. By this I mean that the meaning of public statements was not in their overt content, but communicated the stance of the speaker *vis-à-vis* the authority system. This was true for official spokespersons as for individual citizens. Thus, when the citizen read statements by public officials about the party's commitment to various values or programmes (for example, to working-class leadership, to improved housing policy, to equality between Soviet peoples, or to harmony between man and nature), these statements were understood as largely symbolic and legitimising; they reaffirmed a set of ideological principles linking the present leadership to a past tradition and a present structure of power. At the same time, citizens recognised that the promises contained in the statements were unlikely to be realised in the immediate or near future. Public rhetoric was to serve a legitimising function, which operated, at least passively, in a fairly effective manner, evidenced by the stability of the régime up until 1985. Language was disjoined from reality for the individual citizen as well, for the citizen had to demonstrate compliance or loyalty through language, even though the citizen might believe the opposite. The disjunction between the spoken word and active commitment (both of policy and of individual conscience) underlay the old symbolic politics. While public rhetoric was overtly legitimising, the effect of the disjunction was to create an underlying scepticism toward the veracity of public discourse; this provided the groundwork for the de-legitimisation of public authority once constraints on speech were eased.

Glasnost changed this situation, at least on the surface. Open expression of sentiments was encouraged, in part, as Lapidus suggests, to substitute 'voice for exit' and in part to re-establish the credibility of the leadership and media.[41] What ensued was an unprecedented outburst of public criticism, encompassing virtually all aspects of Soviet life. The ecological area was typical. Glasnost brought vast revelations about the extent, health effects, and costs of environmental degradation. As noted above, the issue took on a symbolic significance: people came to see their own victimisation reflected in the victimisation of guiltless nature. This symbolic importance of the issue was seen in its inclusion in the platforms of a large number of candidates for the Congress of People's Deputies preceding the first broadly contested elections of March 1989 as well as in numerous media reports. Increasingly over time, however, it has become apparent that these verbal commitments are not backed by concrete programmes of action, nor necessarily by a high priority commitment to the issue on the part of the deputy. As more information about environmental problems is released and more intense expressions of official concern emerge both from public officials and from the new legislative bodies, no real change in policy has been forthcoming, nor has any general improvement in the situation ensued. The previous disjunction between public rhetoric and active commitment has reproduced itself, in slightly revised form. The Soviet citizen is frustrated as the hope offered by glasnost has brought little change in real life. This is true in the ecological area as elsewhere. There is an almost uncanny parallel with the earlier 'schizophrenia' of the Brezhnev period – talk and reality seem to be two entirely different things.

At least one dimension of the comparison is even more destabilising, for the new rhetoric allowed by glasnost is debunking and delegitimising. It undermines the old value system without yet producing a positive alternative. A significant portion of the population has reacted with resignation and cynicism as the prospects for real systemic change seem less likely. Furthermore, the disjunction between rhetoric and practice, so characteristic of politics before glasnost, seems to be reincarnated in new forms. Only local victories for environmentalists suggest the possibility of a break-through, but these victories have proven fragile and sometimes ephemeral.

Interest group politics: 'ad hoc' and weakly institutionalised

In the Brezhnev years, interest groups were in some regards highly institutionalised, reflected in Western terms such as 'institutional pluralism'.[42] Most of the 'interest groups' identified by Western scholars were either based in powerful institutional structures (regional party organisations, ministerial structures, large research institutes) or their existence was only inferred from examining statements by scholars and other members of the intelligentsia. The 'groups', for the most part, consisted almost entirely of the political élite and had no corporate existence apart from their association with state or party bureaucratic agencies. At the same time, institutionalised mechanisms for representing the views and interests of the broader masses of the population generally took the form of highly formalised corporatist structures (e.g., Supreme Soviet, trade unions, Komsomol, various official organisations) that were not representative of public opinion.[43] Less formalised means of popular expression had to operate almost entirely underground (e.g., the Helsinki Watch Group). Thus there were few, if any, effective institutional structures for expressing and organising popular demands. In the environmental area, the same pattern applied.

Under perestroika, the old corporatist structures for linking state and society are in the process of delegitimation. They have been superseded by large numbers of *ad hoc* independent public organisations which, as of yet, have not been integrated into the institutionalised structure of conflict resolution. The groups are also very fragmented, despite some efforts to develop umbrella organisations. Thus the mass public still has only weak channels of regularised access to policy-makers and is only indirectly involved in routine processes of conflict resolution. Issues are dealt with as crisis management; in any particular crisis, state and party authorities may be compelled to enter negotiations or hear demands emanating from these new public organisations. But their regularised inclusion in policy deliberation still has not been realised. The allocation of one-third of the seats in the Congress of People's Deputies to official public organisations (trade unions, Academy of Sciences, Komsomol, the CPSU) reflected the failure to make a transition from the old politics of state corporatist representation to a new politics of interest mediation. The decision to eliminate these seats guaranteed to official organisations reflects the continuing delegitimation of the old patterns of representation.

In the environmental area, numerous *ad hoc* public organisations have sprung up all over the USSR. As noted above, they have achieved success in closing plants on an *ad hoc* basis. In the West, environmental activism also gained its strongest initial impetus from such local NIMBY (not-in-my-back-yard) phenomena. In the West, however, there was a whole infrastructure which eased the transition from this kind of fragmented *ad hoc* activism to a more unified and organised movement which could win compromises at the policy level. This infrastructure included a public already oriented toward and schooled in interest group activity, public guidelines for interest group lobbying, competitive party structures, and an array of independent public research institutes and think tanks. In the USSR, this infrastructure is lacking. Thus efforts to unify the environmental movement and to gain access to the policy deliberation process are difficult; that is, the shift from *ad hoc* activism to institutionalised representation has not occurred.

Some evidence of efforts to unify an independent ecological movement can be seen in the formation of umbrella organisations like the Social and Ecological Union or in the formation of a (weak) ecological caucus in the Congress of People's Deputies. The movement is, as noted above, also divided on many issues. Divisions, of course, also characterise Western ecological movements and parties and in themselves need not prevent the institutionalised representation of interests. The greater obstacle is the crisis nature of political decision-making necessitated by the general political and economic crisis in the country. This context inhibits the inclusion of broad-based ecological groups in a policy deliberation process which might explore the deeper trade-offs and values involved in shaping a national environmental policy.

Weak arenas of legitimate conflict mediation

In the environmental area, as in other issue areas, real choices between public goods must be made. For example, introducing a genuine market system could aggravate pollution problems because it would encourage enterprises to externalise costs further in order to increase profit. As in the West, a conflict would emerge over the extent of resources and energy that the state should devote to regulating and controlling this side-effect of a market economy. To some extent, such trade-offs and costs of any given policy choice are inevitable. A legitimate régime will benefit from sufficient good-will on the part of the population to allow acceptance of the costs. This good-will is often based on past policy successes. Or a deliberative decision-making process viewed as legitimate and fair may produce a policy that the majority of the population is willing to accept, for better or worse. Such a deliberative process would presume the inclusion of the affected parties and an open and thorough examination of the cost and benefits of various options. Previous to perestroika, such policy choices were made largely without broad public examination of the alternatives or costs. They were announced as part of a central programme or plan and given formal confirmation by unanimous acclamation of the Supreme Soviet or party bodies. The ambiguities, contradictions, and opportunity costs of policy choices went unexplored and unacknowledged.

In the face of radical problems, rising expectations, and alternative programmes, such top-down decisions no longer have legitimacy, even as the consequences of particular policies may have drastic consequences on the quality of life of the population. Yet the very fundamental choices implied by different policy routes are only beginning to find serious and thorough examination in a legitimate deliberative process. To some extent the Supreme Soviet and/or Congress of People's Deputies provide such a forum, for legislation does meet heated debate there. But examination of that debate indicates that it usually skirts the fundamental value choices – choices are not clearly presented in alternative platforms, fundamental issues are not clearly laid out before the deputies, and the costs of the various alternatives are not addressed head-on. For example, the resolution on the environment adopted by the Supreme Soviet in November 1989 lays out a number of environmental problems and suggests targets for solutions. It indicates, for example, that the safety of nuclear technology should be reviewed.[44] But neither the document nor the published debate indicate any deeper deliberation on the pros and cons of various strategies for future energy policy – a debate which should have included organised anti-nuclear groups as well as ministerial officials, individual deputies and party leaders. Thus the resolution becomes another proclamation which has little legitimacy or authority.

Very few policy choices are cost-free, and public acceptance of the costs requires the existence of a legitimate process of policy deliberation which involves all parties affected by the decision. While some of these kinds of trade-offs are addressed indirectly or in a fragmented fashion under conditions of glasnost, a process of public deliberation and choice has not yet emerged.

Bureaucratic and legislative restructuring

An implicit assumption has long prevailed in the Soviet system that, given appropriate legislation and the establishment of necessary agencies, problems can be solved. This notion continues to hold considerable force under conditions of perestroika. This approach is, of course, basically apolitical, for it substitutes institutional and legislative engineering for a genuine process of conflict resolution. In the environmental area, one can see manifestations of this approach in the past five years: a new environmental agency has been set up, various resolutions and laws have been put in place or promised, but these have not resulted in a real change in the incentive structure or a real resolution to value conflicts. Such legislative and organisational innovations can introduce new actors into the political drama, and they can generate new sources of information and pressure. On the other hand, often such reorganisations and legal enactments make no difference at all because they simply involve the renaming of old agencies or are subject to ineffective implementation. In this case, bureaucratic and legislative restructuring becomes another sort of symbolic politics, and thus may serve to delegitimise public authority rather than enhance it.

Conclusion

Soviet politics is in a state of flux, with a strong popular attraction to particularistic causes and ideologies. Institutionalised mechanisms for mediating fundamental political conflicts are weak, while the attempt to institutionalise new solutions through legislative and bureaucratic restructuring is strong. This pattern is a product of the previous era in which bureaucratic interest group conflict was dominant, but mass involvement occurred primarily through formalised corporatist structures. While these mechanisms seemed to be adequate to maintain political stability during the Brezhnev years, with glasnost their legitimacy was undermined; the result is political instability and a legitimacy gap between public attitudes and policy initiatives. Some new institutional mechanisms such as strong local governments and competing political parties might help to address the problem of effective intermediation of popular demands. However, these responses pose the prospect of simply reinforcing the obsession with particularistic demands. At the same time, central representative bodies are losing popular legitimacy as forums for interest intermediation; particular constituencies have become impatient with ineffective policy proclamations which are seen as insensitive to local needs. In this context a return to authoritarian control may seem attractive to central authorities.

The success of the environmental coalition depends on the success of the now precarious democratic experiment in the USSR. Furthermore, the institutionalisation of political conflict is only a necessary but not a sufficient precondition for successful environmental policy. Effective and legitimate mechanisms of interest representation only set the stage for a national debate; the Soviet population must then confront the brutally difficult choices facing mankind as a whole in the face of the deepening environmental crisis.

Notes

I am grateful to McGill University for funding to support this research, to Marat Khabibullov for assistance in arranging interviews in the USSR, and to Henry Krisch for helpful commentary on the original paper.

1. See Joan DeBardeleben, *The Environment and Marxism-Leninism: The Soviet and East German Experience* (Boulder, CO: Westview Press, 1985).
2. On the parameters of political conflict, see, e.g., Donald Kelley, 'Environmental Policy-Making in the USSR: the role of industrial and environmental interest groups', *Soviet Studies*, 28, no. 4 (October 1976), pp. 570–89.
3. On constraints in the nuclear area, see Joan DeBardeleben, 'Esoteric policy debate: nuclear safety issues in the USSR and GDR', *British Journal of Political Science*, 15 (1985), pp. 227–53.
4. See Boris Komarov (Zeev Wolfson), *The Destruction of Nature in the Soviet Union* (Armonk, NY: M.E. Sharpe, 1980).
5. Kelley, 'Environmental Policy-Making', pp. 574–80.
6. See DeBardeleben, *The Environment and Marxism-Leninism*, Chapter 8.
7. See David R. Marples, *The Social Impact of the Chernobyl Disaster* (London, Edmonton: Macmillan Press, University of Alberta Press, 1988).
8. For example, construction was suspended at the third unit of the Ignalina station in Lithuania; the atomic heat and power station near Minsk will not be built; the Armenian AES has been closed due to local seismic conditions;

the planned Crimean AES will be used for training, and the 4th, 5th, and 6th units of the Southern Ukrainian AES will not be completed. There is now also pressure from the Ukrainian Supreme Soviet to close all of the Chernobyl units presently operating. (See a summary from *Pravda Ukrainy*, 1 March 1990, p. 1, 3, translated in *Current Digest of the Soviet Press*, 42, no. 7 (1990), p. 31.) Likewise, the USSR Supreme Soviet's resolution on Chernobyl calls for the working out of measures to take the Chernobyl units out of operation (*Izvestiya*, 27 April 1990, p. 2).

9. Gosudarstvennyi Komitet po okhrane prirody, *Doklad: Sostoyanie prirodnoi sredy v SSSR v 1988 godu* (Moscow: Goskompríroda, 1989), p. 4.
10. See Joan DeBardeleben, 'Economic and environmental protection in the USSR', *Soviet Geography*, 31 (April 1990), pp. 242–56.
11. Based on interviews with officials of the executive committee of the Moscow City Soviet, 27 April 1990 (Moscow).
12. According to Vasily Pargenov, writing in *Pravda* (21 February 1990, p. 2), more than 1,000 large enterprises have been closed due to pressure from the public on ecological grounds. Particularly affected are the chemical, timber, and pharmaceutical industries, as well as the energy sector. On resultant problems in the energy sector, see the comments by M. Khodzhayev in *Pravda*, 18 February 1990, p. 3. The vice-chairman of the USSR State Statistics Committee, V. Tolkushkin, also noted in his report on economic performance in the first quarter of 1990 that the 'mass shutdown – a public demand' of production facilities which produced hazardous ecological effects was a cause of production shortfalls (*Izvestiya*, 7 April 1990, p. 1). For effects on the production of medicine and newsprint, see Radio Liberty, D. J. Peterson, 'Medicines, newspapers, and protecting the environment', *Report on the USSR*, 2, no. 12 (23 March 1990), pp. 10–13.
13. See, e.g., Kelley, 'Environmental Policy-Making'; Charles Ziegler, 'Issue creation and interest groups in Soviet environmental policy', *Comparative Politics*, 18, no. 2 (1986), pp. 171–92; and Barbara Jancar-Webster, 'Environmental politics in Eastern Europe in the 1980s', in *To Breathe Free*, ed. Joan DeBardeleben (Washington, DC: Wilson Center Press, no. 1, (forthcoming)).
14. See Marat Khabibullov, Joan DeBardeleben, and Arthur Sacks, 'New trends in Soviet environmental policy', *Field Staff Reports*, no. 1 (1990–1); and Kelley, 'Environmental Policy-Making', p. 577. For recent critical commentary on the activities of the Ministry (now transformed into the Ministry of Water Resources Construction) see the letter to *Izvestiya* by People's Deputies S. Zalygin, A. Kazannik, V. Tikhonov, A. Yablokov, and A. Yanshin (*Izvestiya*, 7 February 1990, p. 3).
15. Kelley, 'Environmental Policy-Making', pp. 578–9.
16. For example, see the article on activities in Kirishi by P. C. Filippov, 'Somnitel'noye blago i yavnoye zlo', *EKO*, no. 11 (1988), pp. 138–9.
17. On more critical views of Goskompriroda, its inadequacies are cited as one of the motivations for the formation of the Ukrainian green organisation 'Zeleynyi Svit' in October 1989. See Radio Liberty, *Report on the USSR*, 2, no. 9 (1990), p. 19. The editor of *Sovetskaya Rossiya* warns Goskompriroda that delays in decision-making will work to the harm of the environment in

the controversy over the Bashkir reservoir (*Sovetskaya Rossiya*, 26 April 1990, p. 3). See also the account on a day in the life of the head of Goskompriroda, Nikolai Vorontsov in *Moscow News*, no. 46 (12 November 1989), p. 16. The republican organ of Goskompriroda in Moldavia is also cited for its lack of power or will to act in 'Kuda plesnet "Zelenaya volna"?', *Sotsialisticheskaya industriya*, 11 May 1989.
18. The discussion of the organisation of Goskompriroda is based on interviews at the committee on 26 April 1990, in Moscow.
19. For example, in regard to the Bashkir reservoir, see *Sovetskaya Rossiya*, 26 April 1989, p. 3.
20. *EKO*, no. 11 (1988), pp. 130–1.
21. See *Moscow News*, no. 42 (23–30 October 1988), p. 4; no. 17 (30 April–7 May 1989, p. 14); and no. 20 (21–28 May 1989), p. 14.
22. S. Sokolov, 'Prirodovol'tsy', *Komsomol'skaya pravda*, 25 December 1988.
23. Vladimir Prokushev, 'Kruglyi stol' . . . Samoubiits?' *Zhurnalist*, 10 October 1988, pp. 4–7, 11.
24. *Trud*, 6 May 1989, p. 3. The agreement by the Leningrad city executive committee to participate in a project with American investors to produce a tourist and sports centre on the northern shore of the Bay of Finland was apparently largely motivated by the city's need for foreign currency. On the project and public protests to it, see *Sovetskaya Rossiya*, 25 February 1989, p. 3.
25. Prokushev; and 'Chto vozrozhdayet "vozrozhdenie"', *Sovetskii zhurnalist*, 12 December 1988.
26. On the investigatory expedition sponsored by *Yunost'*, see its issue no. 2, 1989; on *Novyi mir*'s co-sponsorship of an expedition to the Aral Sea by scientists and journalists see *Moscow News*, no. 36 (11–18 Sept. 1988), p. 7. *Sovetskaya Rossiya*, along with editors of other newspapers, was instrumental in the formation of the Public Committee to Save the Aral (*Sovetskaya Rossiya*, 5 February 1989, p. 3). *Trud* designated itself an intermediary for the creation of a committee on the future of Gornyi Altai (6 May 1989, p. 3).
27. See, for example, speeches by candidates in Kazan', published in *Vechernaya Kazan'* on 11 March 1989, pp. 2–4; and 29 April 1989, pp. 2–3. Only five of the twenty statements published by candidates in these issues do not include mention of ecological matters. Of major importance was the controversy over construction of the Tatar Atomic Energy Station, a project eliciting much popular opposition in Kazan'. (I am grateful to Marat Khabibullov, of Kazan' State University, for the material from *Vechernaya Kazan'*).
28. An interview with Aleksei Yablokov, on 24 April 1990 (Moscow) was helpful to the author in writing this article.
29. Information in this paragraph is based on an interview with Mikhail Viktorov, the leading specialist for societal relations at Goskompriroda, 26 April 1990 (Moscow); and Ruben A. Mnatsakanyan, researcher in the Faculty of Geography at Moscow State University (Harrogate, England, 25 July 1990).
30. Interviews with Viktorov and Mnatsakanyan; and see, for example, the

article 'Kuda plesnet "Zelenaya volna"?' in *Sotsialisticheskaya industriya*, 11 May 1989, describing the formation of the ecological organisation 'Aktsiunia vedre' in Moldavia. The article underlines the important role of scientists and journalists but also the involvement of workers and students. Numerous other articles or new ecological initiatives highlight the role of scientists in providing expert substantiation for claims of environmentalists and the role of journalists and members of the creative intelligentsia in publicising demands.

31. See Theodore H. Friedgut, 'Ecological Factors in the July 1989 Mine Strike', *Environmental Policy Review*, 4, no. 1 (January 1990), pp. 53–8.
32. See, e.g., the situation in Kirishi, noted in Filippov, 'Somnitel' noye blago i yavnoye zlo', p. 131.
33. For example in Ryazan (Sokolov, *Komsomol'skaya pravda*, 25 December 1988).
34. For example, in Moldavia, as discussed in *Sotsialisticheskaya industriya*, 11 May 1989.
35. The information here on the Environmental Business Association comes from an interview with its representative, Nick. N. Totsky, 26 April 1990 (Moscow).
36. See DeBardeleben, 'Economic reform', pp. 245–6.
37. *Yunost'*, no. 2, 1989.
38. See Ronald Ingelhart, *The Silent Revolution* (Princeton: Princeton University Press, 1977); and Philip Lowe, 'Environmental management in West European countries: social movements, ecological problems, and institutional responses', in DeBardeleben (ed.), *To Breathe Free* (forthcoming).
39. See the 'Panel on nationalism in the USSR: environmental and territorial aspects', *Soviet Geography*, 30 (June 1989), pp. 471–84.
40. For example, see the campaign platform of the Bloc of Russian Public-Patriotic Movements published in *Literaturnaya Rossiya*, 29 December 1989, trans. in *CDSP*, 42, no. 1 (1990), pp. 1–4, 23. The platform includes an attack on the All-Union departments and demands full control for Russia over its natural resources.
41. Gail Lapidus, 'State and society: toward the emergence of civil society in the Soviet Union', in Seweryn Bialer (ed.), *Politics, Society, and Nationality inside Gorbachev's Russia* (Boulder, CO: Westview, 1989), p. 123.
42. Jerry Hough, 'The Soviet Union: petrification or pluralism', in Hough (ed.), *The Soviet Union and Social Science Theory* (Cambridge, MA: Harvard University Press, 1977), pp. 22–34. For a broader discussion of institutionalisation in the Brezhnev years and how it relates to present reform efforts, see Peter Hauslohner, 'Politics before Gorbachev: de-Stalinisation and the Roots of Reform', in Bialer (ed.), *Politics, Society, and Nationality*, pp. 41–90.
43. For a corporatist interpretation of Soviet politics under Brezhnev, see Valerie Bunce, 'The political economy of the Brezhnev era: the rise and fall of corporatism', *British Journal of Political Science*, 13 (April 1983), pp. 129–58. For an application of corporatism to the environmental area, see Ziegler, 'Issue creation', pp. 171–92.
44. *Izvestiya*, 3 December 1989, p. 3.

7 Water management in Soviet Central Asia: problems and prospects

Philip P. Micklin

An abundant and assured supply of fresh water is essential to modern societies. Water withdrawals are particularly large in arid regions of extensive irrigation development. But irrigation consumes a significant proportion of water withdrawn and therefore depletes river flow and ground water reservoirs. The development of rational water management policies and approaches for arid regions with large-scale irrigation, including balancing competing economic and ecological demands, is not only technically complicated but fraught with social, economic, and political difficulties.

The most serious water management problems in the USSR occur in Central Asia, a region situated in the most arid zone of the country. Intensive development of irrigation here has depleted river flow and led to the drying of the Aral Sea, a huge saline lake, with accompanying severe impacts. Until the mid-1980s, the Soviet government intended to alleviate the region's water management troubles by large-scale, long-distance water importation from Siberian rivers far to the north. But this project has been halted and local means of resolving water supply problems are being pursued. These, however, may fail to adequately resolve the water crisis.

Water resources and usage

Central Asia lies among the deserts of the extreme southern part of the Soviet Union, with the Caspian Sea to the west and high mountain chains to the south and east (figure 7.1). Central Asia is defined in this paper to include not only the Uzbek, Tadzhik, Kirgiz, and Turkmen Soviet Socialist Republics (SSRs) but the two southern oblasts of the Kazakh SSR (Kzyl-Orda and Chimkent), also found in the Aral Sea drainage basin and having a commonalty of water management problems with the four republics (see figure 7.1). Central Asia's population was 35 million on 12 January 1989: 12 per cent of the USSR total. The rate of natural increase (births minus deaths) averaged 2.54 per cent

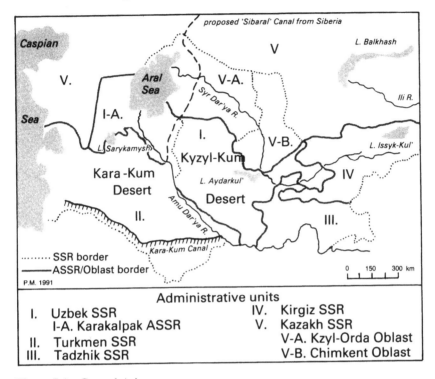

Figure 7.1 Central Asia

over the intercensul period 1979–89 compared to a national rate of 0.87 per cent (table 7.1). This is the most rapid growth of any region in the USSR and exceeds the rates in many developing countries. High fertility, the young age structure, and minimal out-migration make rapid population growth for Central Asia a near certainty well into the next century. Optimistically assuming a slowing of the average growth rate to 2 per cent for the period 1989–2000 and to 1.5 per cent for 2001–2010, population still would grow to 44 million, a 24 per cent increase, by the turn of the century, and to 50 million, a 42 per cent increase, by 2010.

Thermal conditions for plant growth are the best in the USSR.[1] Hence, cultivation of heat-loving crops such as grain corn, sorghum, rice, and soy is possible over much of the region and cotton can be grown in the desert plains and foothills in all but its northern part.

The surrounding mountains capture plentiful moisture and store it in snow fields and glaciers whose run-off, primarily during spring thaw, feeds the region's rivers. Estimated average annual river flow in Central Asia is 122 cubic kilometres per year. The Aral Sea drainage basin,

Table 7.1. *Central Asia: area and population characteristics*

Administrative unit	Area (000's of sq. km.)	Population on 1/12/89 (millions)	% total pop.	% average annual natural increase, 1979–1989
Uzbek SSR	447.4	19.91	56.37	2.57
Kazakh SSR*	344.4	2.48	7.02	1.54
(Kzyl-Orda)	(228.1)	(0.65)	(1.84)	(1.41)
(Chimkent)	(116.3)	(1.83)	(5.18)	(1.58)
Tadzhik SSR	143.1	5.11	14.47	2.96
Kirgiz SSR	198.5	4.29	12.15	2.01
Turkmen SSR	488.1	3.53	9.99	2.49
Total	1,621.5	35.32	100.00	2.54†

* Kzyl-Orda and Chimkent *oblasts* only.
† This is a weighted average.
Source: compiled from *Narodnoye khozyaistvo SSSR za 70 let* (Moscow: Finansy i statistika, 1987), pp. 393–4 and *Pravda Vostoka*, 29 April 1989, p. 2.

which includes the two largest rivers in the region (Amu Dar'ya, 73 cubic kilometres per year, and Syr Dar'ya, 37 cubic kilometres per year) accounts for 90 per cent (110 cubic kilometres) of this. Discharge is maximum where rivers leave the mountains but decreases rapidly as they cross the deserts. The Amu and Syr Dar'ya (until the 1960s) lost around half their flow before reaching the Aral Sea. Usable supplies of ground water (i.e., ground water that could be withdrawn on a sustained basis without adversely affecting surface flow) are estimated at 18 cubic kilometres a year. Thus, aggregate average annual water resources for Central Asia are around 140 cubic kilometres a year, nearly all of it found in the Aral Sea drainage basin.

Water use in Central Asia is enormous. For 1980, withdrawals were estimated at 134 cubic kilometres with consumptive use (water directly lost to evaporation and transpiration or incorporated into plants, animals, or other products) of 80 cubic kilometres (60 per cent of withdrawals).[2] The balance of 54 cubic kilometres (40 per cent of withdrawals) constituted return flows from sources such as leakage from canals and pipes, surface and ground water run-off from water applications, water collected by irrigation drainage systems, and end discharges of canals or pipes. A significant portion of return flows, primarily from water used for irrigation, is subsequently evaporated or transpired (i.e., consumptively used) rather than returning to rivers or adding to the ground water reservoir. Central Asia accounted for 39 per cent of withdrawals and 49 per cent of consumptive use nationally in 1980. Consumptive use and reservoir evaporation (together totalling 91 cubic kilometres) were 74 per cent of average annual flow and 65 per cent of average annual water resources in Central Asia.

More recent information is available for the Aral Sea basin which accounted for 98 per cent of withdrawals and 99 per cent of consumptive use in Central Asia in 1980 (table 7.2). These data indicate that in 1988 water withdrawals for all purposes were 167 cubic kilometres, 124 per cent of average annual water resources and 160 per cent of average annual flow. Withdrawals were more than 100 per cent of the resource because return flows are repetitively used downstream. Consumptive use plus water losses in transportation totalled around 117 cubic kilometres, 87 per cent of average annual water resources and 99 per cent of average annual flow. The year 1988 was an above average flow year; nevertheless, these data illustrate that the surface water resources of the Aral Sea basin are essentially exhausted. During the 1980s, excluding the heavy flow years 1987 and 1988, little water reached the Aral Sea. Both the Amu and Syr Dar'ya basins are in dire straits with the situation somewhat more serious for the latter than the former. The apparent major increase in withdrawals (from 134 to 167 cubic kilometres) between 1980 and 1988 is probably an artifact of the differing estimation techniques used to create the two data sets rather than a reflection of a real increase in water withdrawals, which are likely to have held fairly stable through the 1980s.

Irrigation is the dominant user of water in Central Asia, accounting in 1980 for 90 per cent of withdrawals, 95 per cent of consumptive use (excluding reservoir evaporation), and 84 per cent of return flows.[3] The fundamental cause of the water crisis in Central Asia is irrigation. Water use for other purposes is, by comparison, small. Even minor improvements in irrigation water-use efficiency could free sufficient water to meet future needs of other economic and social sectors in Central Asia.

Central Asia is the most important zone of irrigation in the USSR. Irrigation has been practised there for thousands of years. By 1990, the area irrigated by state-run organisations was nearly 8 million hectares, accounting for around 40 per cent of the national total.[4] The USSR is the world's third largest cotton producer. All Soviet cotton is irrigated and 95 per cent is raised in Central Asia. This is the region's pre-eminent crop. Irrigation is also crucial to food and fodder production: 40 per cent of the USSR's rice, one-third of its fruit and grapes, and a quarter of its vegetables and melons are grown on irrigated lands here.

Table 7.2. Water resources and use in the Aral Sea basin in 1988

Water resource/use parameter	Amu Dar'ya basin	Syr Dar'ya basin	Other*	Aral Sea basin
Average annual flow (km³)	69.50	37.00	11.05	117.55
Confirmed safe groundwater yield (km³)	9.00	7.40	n.a.	16.40
Total average annual water resources	78.50	44.40	n.a.	133.95
Water withdrawals (km³)	77.35	64.22	25.71	167.28
as % of average annual water resources	98.53	144.64	n.a.	124.88
surface sources	75.92	60.30	23.47	159.69
as % of withdrawals	98.15	93.89	91.27	95.46
as % of average annual flow	111.29	173.57	232.69	142.31
groundwater	1.43	3.92	2.24	7.59
as % of withdrawals	1.85	6.11	8.73	4.54
as % of groundwater yield	20.51	82.53	n.a.	27.68
Consumptive water use (km³)	57.07	43.36	20.56	120.98
municipal	0.99	1.85	0.77	3.61
as % of use	1.74	4.26	3.75	2.98
production (without agriculture)	3.49	5.52	3.21	12.22
as % of use	6.11	12.73	15.63	10.10
agricultural	52.59	35.99	16.57	105.15
as % of use	92.15	83.00	80.62	86.91

Water loss in transport (km³)	18.49	10.45	4.70	33.64
as % of withdrawals	23.90	16.28	18.28	20.11
as % of average annual flow	26.60	28.25	42.52	28.62
as % of average annual water resources	23.55	23.54	n.a.	25.11
Return flows to surface waters (km³)	16.95	15.89	5.05	37.89
as % of withdrawals	21.91	24.74	19.63	22.65
as % of average annual flow	24.39	42.95	45.67	32.23
as % of average annual water resources	21.59	35.79	n.a.	28.28
Polluted return flows (km³)	0.09	0.36	0.12	0.57
Consumptive use plus transport losses (km³)	58.60	37.92	20.21	116.73
as % of withdrawals	75.77	59.05	78.61	69.78
as % of average annual flow	84.32	102.49	182.90	99.31
as % of average annual water resources	74.65	85.41	n.a.	87.15

* Basins of the Kashkadar'ya, Zeravshan, rivers of Afghanistan and Turkmenistan (including Kara-Kum Canal).

n.a.: not available.

Sources: Taken or calculated from *Okhrana okruzhayushchei sredy i ratsional'noye ispolzovaniye prirodnykh resursov v SSSR* (Moscow: Goskomstat, 1989), pp. 66–7, and other sources.

Irrigation provides over 90 per cent of crop production in Central Asia.[5]

Given the near certainty of continued rapid population growth in Central Asia for the foreseeable future, rapid expansion of the regional economy and of food production is essential. Already per capita consumption in a number of basic foodstuffs is significantly below national averages.[6] Irrigation as a means to meet Central Asia's growing employment and food needs, has much to recommend it. It is labour intensive and does not require a highly skilled work force, which is also true of ancillary industries such as food processing and cotton textile and clothing manufacture. Crops are grown without irrigation on the more humid slopes of the region's foothills and mountains and hardy livestock (sheep, goats, and camels) are raised on natural desert vegetation. But crop yields and meat production are much lower and more variable. Until the mid 1980s, continued growth of irrigation in Central Asia was assumed. The long-range reclamation programme approved in 1984 projected the irrigated area in Uzbekistan, Turkmenia, Kirgizia, and Tadzhikistan to rise 23–7 per cent between 1985 and 2000.[7] Because of greater emphasis by the Gorbachev régime on alternative means of increasing agricultural production and, particularly, the dire water supply situation in Central Asia, planned irrigation expansion here has been scaled-back.[8]

Many Soviet water management experts contend the water resources of the Aral Sea basin reached full utilisation in the early 1980s owing to heavy withdrawals for irrigation. An authoritative 1987 study of water management in the USSR concluded that total withdrawals for irrigation were around 100 cubic kilometres in the basins of the Amu and Syr Dar'ya, with 35 cubic kilometres lost to filtration.[9] Of total filtration losses, 14 cubic kilometres were ultimately used consumptively, suggesting 21 cubic kilometres (35–14), or 21 per cent of withdrawals, constituted water returned to rivers and 79 cubic kilometres or 79 per cent overall consumptive use. The Syr and Amu Dar'ya are considered to have the most strained water balances of the USSR's major rivers.

Water management improvement measures

Water management in general and irrigation in particular in Central Asia have become the focus of sharp controversy. The critics contend that water wastage in irrigation here is enormous and maintain that, with moderate effort, sufficient amounts of water can be freed (35–50 cubic kilometres) to meet regional needs far into the future.[10] Water management specialists, on the other hand, are more cautious, predicting the amount of water that can be freed is much less than the optimists believe (10–20 cubic kilometres), that it could somewhat alleviate but by no means 'solve' regional water problems, and that implementing the necessary water saving measures will be lengthy, costly, and complicated.[11]

There is general agreement that, first and foremost, water use efficiency in irrigation must be improved. A rough estimate for the Aral Sea basin, which has some 96 per cent of the irrigated area in Central Asia, is that by the early 1980s only 60 per cent of withdrawals arrived at the field with 40 per cent (perhaps equalling 50 cubic kilometres) lost in the conveyance system.[12] A more accurate measure of water-use efficiency in irrigation is the portion used productively by crops (i.e., necessary for growth and survival). This figure may have been as low as 44 per cent in the late 1970s in the basins of the Amu and Syr Dar'ya. However, not all of the residual 56 per cent was lost for further use: one quarter to one third constituted return flows which re-entered the river network by surface or ground water routes. Nevertheless, Central Asian irrigation in the early 1980s provided major opportunities for improvement and water savings.

Soviet water management experts have set the average efficiency target for irrigation systems in Central Asia around 80 per cent. This refers to the percentage of withdrawn water arriving at field side. Assuming 1980 withdrawals for irrigation were 120 cubic kilometres, this one-third efficiency improvement would have allowed irrigating the same area with 30 cubic kilometre lower withdrawals (with 60 per cent efficiency, 72 cubic kilometres of the 120 arrives at the field; to obtain the same delivery, 72 cubic kilometres, with 80 per cent efficiency requires withdrawal of only 90 cubic kilometres).[13] Again, this does not imply a net savings of 30 cubic kilometres since irrigation return flows to rivers would be substantially reduced.

The improvement of irrigation efficiency in Central Asia has been a priority since 1981 when water scarcities caused strict limits to be placed on water consumption.[14] It has received additional emphasis under Gorbachev as part of the general programme for agricultural intensification and was assigned the highest priority by the Ministry of Reclamation and Water Management (renamed the Ministry of Water Management Construction in 1989 and removed from ministerial status in 1990) for the 12th five-year plan.[15] The effort to raise irrigation efficiency has fortunately had results in Central Asia. Data from the Central Asian Institute for Irrigation Research (SANIIRI) show average per hectare withdrawals falling from 18,700 cubic metres per hectare in 1980 to 13,700 cubic metres per hectare in 1986.[16] This allowed a 15 per cent increase in the irrigated area over this period with a 16 per cent decrease in water withdrawals. However, 1986 was an especially water-short year which necessitated abnormally tight restrictions on water use. Using it as a comparison may overestimate efficiency improvements.

There are a variety of measures and strategies to improve the water-use situation in Central Asia. Rebuilding of older irrigation systems to reduce water losses has received the greatest attention. Given the long history of irrigation development here, a larger share of the irrigated area is served by antiquated and inefficient irrigation systems than in any other part of the USSR. Thus, 70 per cent of new irrigation investment here during the 12th five-year plan was to be devoted to renovation.[17] By AD 2000, irrigation systems on 3.3 million hectares are to be reconstructed.[18] According to Polad-Polad Zade, First Deputy Minister of the former Ministry of Water Management and Reclamation, this could cost 25 billion roubles and free only 10 cubic kilometres of water annually.[19]

Reconstruction involves implementation of a variety of measures.[20] Fundamentally important are water-loss reductions from irrigation canals by lining them with concrete or other coverings such as plastic sheeting or clay, and by shortening their aggregate length through consolidation. Installing or improving collector-drainage facilities to remove excess ground water and prevent water logging and soil salinisation is also crucial. Most older irrigation systems in Central Asia either lack engineered drainage networks entirely or have crude open channels, frequently choked with weeds, that are not effective. Proper drainage not only prevents water logging and soil salinisation but also lowers water usage by greatly reducing the amount of water used during the late winter–early spring period of soil flushing, intended to remove accumulated salts before the new growing season.

Improving water application at the field is another essential measure for raising the efficiency of water use. In the early 1980s, 98 per cent of irrigation in Central Asia was by surface methods, where water flows from a canal or flume directly onto the fields and then by furrows to the crops.[21] The efficiency of furrow irrigation is inherently low but can be raised substantially through the automation and mechanisation of water deliveries, the use of siphons, hoses, and movable and rigid pipes to control more precisely the delivery of water to the crops and the levelling of fields to equalise water distribution. There are also more modern and efficient irrigation technologies such as sprinkling, drip, and inter-soil methods whose use could be expanded to some degree in Central Asia, though each has limitations (e.g., expense, energy consumption, suitability for only certain crops, low efficiencies in these desert conditions). Consequently, surface irrigation will remain the most important water application technology here.

Automation, computerisation, and tele-mechanisation of large irrigation systems is being implemented to improve water usage. The basis of the programme is the installation of water measurement and regulating devices along main and distributory canals that can transmit data to and be directed by a central facility. These technological innovations provided the basis for the establishment of centralised water management authorities (basin water management associations) for the Amu Dar'ya and Syr Dar'ya in 1988.[22] Among their other duties, these have responsibility for inter-republic and inter-sector allocation of river flow and technical operation of all water withdrawal facilities and pumping stations. The systems are supposed to gather and process information on estimated water resources and water needs within the basins, determine an allocation plan for water use for a year, control the distribution of water within and among republics, and keep track of actual water allocations, uses, and conditions.

Accompanying the change of title of the Ministry of Reclamation and Water Management (Minvodkhoz) to the Ministry of Water Construction (Minvodstroi) in 1988 was a diminution of its ability to control water use and allocation in Central Asia. Its authority was removed to enforce the limits on water withdrawals established annually by Gosplan.[23] If these are exceeded, the matter must now be referred back to Gosplan in Moscow and in complicated cases even to the Council of Ministers of the USSR. Republics are also asserting the right to control water usage within their territory through the Councils of Peoples' Deputies. According to water management experts, the lack of an independent state organisation with power to regulate water use in Central Asia is contributing to excessive withdrawals and wasteful use as each republic attempts to manage its own water resources without taking account of the interests of its downstream neighbours. This has also led to water disputes between republics.[24] With the dissolution of Minvodstroi, the situation will likely grow worse.

Another water conservation measure being implemented is a shift from high water-consuming crops such as cotton and rice to lower users such as vegetables. This change will not only lower per hectare withdrawals but contribute to the improvement of regional food supplies.[25] Nevertheless, cotton will continue to be the dominant irrigated crop in Central Asia for the foreseeable future because of existing infra-structural inertia and its great economic importance to the region. It has also been proposed to improve the meat supply by shifting from cotton to fodder crops such as alfalfa or corn. These crops, however, depending on how they are grown, can require more water than cotton.[26] Efforts are being made to raise the yield per cubic metre of applied water of all crops and to refine irrigation water application standards to adjust them more precisely to crop water-consumption requirements.[27]

There are opportunities to develop new or currently under-utilised water resources for irrigation in Central Asia. Ground-water usage could perhaps be expanded to 18 cubic kilometres a year without adverse effects on surface water flow.[28] A larger share of irrigation drainage-water might be reused but much of it is so saline that without dilution it would damage most crops.[29] The reservoir system along the Amu and Syr Dar'ya, when finished, will allow the increase of available water resources from these rivers, compared to natural conditions, by 15 cubic kilometres during low-flow years.[30] On the other hand, as well as having adverse ecological effects, reservoirs lose large amounts of water by evaporation, transpiration from phreatophytes (water-loving plants) growing in shallows or along banks, and exfiltration through their sides and bottoms. Finally, it is possible to utilise periodic run-off collecting in ephemeral streams and clay-pan basins for small-scale irrigation.

On the economic and institutional side, water pricing is the most promising means of improving water use in irrigation. Water used for agricultural purposes has been provided free which promotes inefficiency and waste. Although charges for irrigation water have been employed at times in the Soviet Union, the system now under discussion goes far beyond past measures.[31] It would force state and collective farms to pay for water delivered to them by republic water management agencies, who in turn are to be charged for their water withdrawals from rivers and ground water. Although introduction of a price for irrigation water in principle would be beneficial, implementation is beset by problems such as a lack of water-delivery measurement facilities and, most importantly, the out-dated and illogical system of agricultural prices. Consequently, irrigation-water pricing was introduced on an experimental basis in only three oblasts, including Tashkent in Uzbekistan, during 1988–9. Based on this experience, it is planned to begin implementing irrigation water charges nationally in 1991.[32]

The Aral Sea problem

If the water management difficulties of Central Asia related to irrigation were not enough, there is the added complication of the desiccation of the Aral Sea. Situated in the heart of Central Asia, this huge, shallow, saline lake has no outflow; its level is determined by the balance

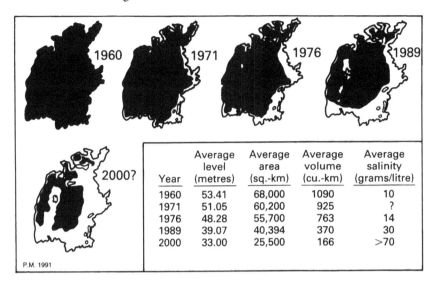

Figure 7.2 The changing profile of the Aral Sea

between river and ground water inflow and precipitation on its surface on the one hand, and evaporation from the sea on the other. The sea, during its 'modern' geological history, covering the last 10–20,000 years, has experienced major recessions and advances.[33] Over most of this period, these have been due to natural factors, but during the last several millennia human activities, at times, have played a key role in these fluctuations. For the period from 1910, when accurate and regular level observations began, to 1960, the Aral was stable with the range of level changes less than one metre. However, during the past thirty years the sea's surface has dropped precipitously (figure 7.2). In 1960, sea level was at 53.4 metres, area was 68,000 square kilometres, volume 1090 cubic kilometres, and average salinity 10 grams per litre.[34] The Aral was the world's fourth largest lake in area. By 1989, sea level had fallen over 14 metres, area had decreased by 40 per cent and volume by two-thirds, and average salinity had risen to 30 g/l.[35] The sea had dropped to sixth place in area among the world's lakes.

This recent recession and its associated impacts have been the most rapid and pronounced in several thousand years. The severe and widespread ecological, economic, and social consequences are steadily worsening. The Aral situation has been characterised in the Soviet press as 'one of the very greatest ecological problems of our century', and compared in magnitude to the Chernobyl nuclear accident and the Armenian earthquake.[36]

As in the past, the cause of the Aral's modern recession is marked diminution of inflow from the Syr and Amu Dar'ya, the sea's sole sources of surface inflow, which has increasingly shifted the water balance toward the negative side. Excepting the heavy-flow year of 1969, the trend of river discharge to the sea has been steadily downward since 1960.[37] Evaporative losses from a shrinking water body diminish as its area decreases, forcing the water balance toward equilibrium. Hence, in the future, the Aral's level, supported by residual irrigation drainage and ground-water inflow, should stabilise. However, this is not likely to occur for many years since the difference between inflow and net evaporation (evaporation minus precipitation) is currently large and negative. If present processes are allowed to continue unchecked, the Aral will be reduced to several briny remnants in the next century.

The causes of reduced inflow since 1960 are both climatic and anthropogenic. A series of dry, naturally low-flow years occurred in the 1970s and 1980s.[38] But the main factor reducing river flow has been large consumptive withdrawals, mostly for irrigation. Although irrigation has been practised in the basins of the Amu and Syr Dar'ya for millennia, the consumptive use of water by irrigation prior to the 1960s did not measurably reduce inflow to the Aral. This usage was nearly compensated for by correspondingly large reductions of natural losses from evaporation, transpiration and filtration, particularly in the deltas of the Syr and Amu Dar'ya, owing primarily to truncated spring flooding. Also, the installation of drainage networks increased irrigation return flows to these rivers, albeit of reduced quality. Thus, almost the same amount of water reached the sea.

Factors that compensated for the earlier growth of consumptive withdrawals reached their limits in the 1960s. Hence, as irrigation in the Aral Sea's basin expanded from around 5 to 7.6 million hectares over the past three decades, consumptive water use from this doubled from 40 to over 80 cubic kilometres (and perhaps to as much as 100 cubic kilometres), but the increase in water usage has not been balanced by commensurate reductions in natural losses.[39] The disproportionate increase of consumptive use compared to the growth of the irrigated area resulted from the development of new areas – such as the Golodnaya (Hungry) steppe along the Syr Dar'ya, where huge volumes of water went to saturate dry soils and drainage water flowed into the desert or natural depressions to evaporate rather than returning to rivers – and from increased soil flushing needed to combat secondary salinisation. The creation of new reservoirs also contributed to losses of river-flow owing to the filling of dead storage and increased evaporation from their surfaces.

The Kara-Kum Canal is the single most important irrigation and water-supply system reducing inflow to the Aral in recent decades. The largest and longest irrigation canal in the USSR, it stretches 1,300 kilometres westward through the Turkmen Republic along the southern margins of the Kara-Kum desert from where the Amu Dar'ya emerges from the mountains (see figure 7.1). Between 1956 and 1987, 236 cubic kilometres of water were diverted into it from the river as annual withdrawals rose from less than one to 12 cubic kilometres.[40] All of the water sent along the Kara-Kum Canal is lost to the Aral. Nevertheless, diversions into the Kara-Kum from 1956 through 1986 were only 15 per cent of the inflow necessary to maintain sea level at the 1960 mark (53 metres) and in recent years it has probably accounted for not more than 10–12 per cent of aggregate consumptive water use in the Aral Sea basin. Thus, as water management experts from the Turkmen Republic have rightly claimed, it is simplistic and unfair to lay the basic blame for the Aral's recession on the Kara-Kum Canal.

The Aral: environmental and ecological consequences

When plans for a major expansion of irrigation in the Aral Sea basin were developed in the 1950s and 1960s, it was anticipated that this would reduce inflow to the sea, substantially reducing its size. At the time, a number of water management and desert development experts believed this a worthwhile trade-off: a cubic metre of river water used for irrigation, they calculated, would be more economically beneficial than the same volume delivered to the Aral Sea.[41] They based this calculation on a simple comparison of estimated benefits from the sea (e.g. fisheries and transportation). Although a small group of scientists warned of serious negative effects from the Aral's desiccation as early as the mid-1960s, they were not heeded.[42] Time has proven the more cautious scientists not only correct but conservative in their predictions. Several of the most important impacts are briefly discussed below.

The Aral contained an estimated 10 billion metric tons of dissolved salts in 1960.[43] As the sea has shrunk, enormous quantities of salts have deposited on the former bottom, which attained an area of nearly 28,000 square kilometres by 1989. Owing to the salt concentrations, the dried bottom is proving stubbornly resistant to natural and artificial re-vegetation. Consequently, the airborne transport of salt and dust has become a severe problem. The largest plumes arise from the exposed strip (up to 100 kilometres wide) along the sea's north-eastern and eastern coast. The lighter aerosols entrained in them are lifted up to 4 kilometres and deposited as far as 400 kilometres downwind.[44] Major dust-salt storms were first spotted by Soviet cosmonauts in 1975 and, by 1981, twenty-nine large storms had been identified by Soviet scientists from analysis of satellite imagery.[45] The majority of storms moved in a south-west direction which carried them over the ecologically and agriculturally important delta of the Amu Dar'ya. More recent observations by Soviet cosmonauts indicate that the frequency and magnitude of the storms is growing as the Aral shrinks.[46]

There is dispute as to how much salt, sand, and dust is blown from the Aral's dried bottom. One source estimates the total volume of sand, dust and salt at 140 million metric tons.[47] Another sets the amount of salt at 43 million tons.[48] Field measurements between 1981 and 1986 indicated dust and salt deposition at no more than 9.5 tons per hectare – less than half of earlier estimates.[49] The salts, deposited as aerosols by rain and dew, are toxic to plants – especially sodium chloride and sodium sulphate – harmful to animals who ingest them when grazing, and even cause electrical shorting of power lines leading to fires when they are deposited on insulators.[50] The blowing dust and salt is also believed to cause serious respiratory problems and illnesses for people and animals.[51]

As the sea has shallowed, shrunk, and salinised, aquatic productivity has rapidly declined. By the early 1980s, native fish species had disappeared, although a few introduced salt-tolerant types remained.[52] The catch of fish which reached 44,000 metric tons in the 1950s fell to zero. Major fish canneries at the former ports of Aral'sk and Muynak have slashed their work force and barely survive on the processing of high-cost fish from distant oceans. Employment directly and indirectly related to the Aral fishery, reportedly 60,000 in the 1950s, has disappeared. The demise of commercial fishing and other adversities have led to an exodus from Aral'sk and Muynak and the abandonment of former fishing villages.

The Aral's shrinkage and the greatly reduced flow of the Syr and Amu Dar'ya has devastated these rivers' deltas.[53] Thirty years ago, the deltas not only possessed great ecological value but provided livestock pasturage, spawning-grounds for commercial fish, reeds used in the paper-making industry and for home construction by local inhabitants, and opportunities for commercial hunting and trapping. These uses have been lost or severely degraded. For example, the area of *Tugay* forests, composed of dense stands of water-loving plants (phreatophytes) mixed with shrubs and tall grasses fringing delta arms and channels to a depth of several kilometres, had been halved by 1980. Disappearance and degradation of vegetational complexes and water table drops have contributed to desertification in both deltas. The formerly rich deltaic fauna, including some endangered species of waterfowl, has been severely depressed. For example, the number of major nesting bird species in the Syr Dar'ya delta has dwindled from 173 to 38.[54]

Several other major adverse consequences of the Aral's recession are also apparent.[55] Climate around the Aral has become more continental and dryer, with warmer summers, cooler winters and lowered humidity, particularly within 50 to 60 kilometres of the former shoreline.[56] Allegedly, this has shortened the growing season on the northern margins of cotton raising in the Karakalpak ASSR sufficiently to force a switch to rice cultivation. The flow of artesian wells and level of ground water has dropped all around the sea, leading to dried wells and springs and the degradation of natural plant communities, pastures and hay fields. The reduction of river flow, the salinisation and pollution by agricultural, industrial, and urban effluents of remaining in-stream discharge, and the lowering of ground water levels have caused drinking water supply problems. Drinking water contamination is believed the main cause of high rates of intestinal illnesses, hepatitis, kidney failure and liver ailments, oesophageal cancer, birth defects and even typhoid and cholera.[57] The rate of oesophageal cancer in Muynak, the former sea port in the Amu Dar'ya delta, is 15 times the national average.[58] Desert animals drinking from the Aral Sea are also dying because of its greatly increased mineralisation, including the endangered *kulan* (Asiatic wild ass) and *saiga* (steppe antelope) that live in the *zapovednik* on Barsakel'mes Island.[59] Poor quality drinking water, generally poor health care, inadequate diet, and frequent child-bearing contributed to an infant mortality rate of 60 per 1,000 in the Karakalpak ASSR in 1989.[60] Tashauz oblast of Turkmenia, partially situated in the Amu Dar'ya delta, had an infant mortality rate of 75 per 1000 in 1988.[61] The average rate for the USSR and the US, respectively, for these same years was 25 and 10.

There are no accurate figures on damages associated with the Aral's recession, but a 1979 study concluded that aggregate damages within the Uzbek Republic totalled 5.4 to 5.7 billion roubles.[62] Two water management experts have cited 100 million roubles per year as the 'social product' losses in the Amu Dar'ya delta.[63] A popular article listed, without elaborating, 1.5 to 2 billion roubles as the annual losses for the entire Aral Sea region.[64]

The Aral Sea: ameliorative measures

If measures are not taken to deliver more water to the Aral, it will continue to shrink. By 2000, it could easily fall an additional 6 metres from its 1989 level and lose another 15,000 square kilometres of surface (figure 7.2). Early in the next century, it would become a set of lifeless, residual brine lakes. Restoring the sea to its 1960 dimensions would take an average annual inflow of around 55 cubic kilometres, over seven times the estimated average yearly inflow for 1980–89 of 7.5 cubic kilometres, and require many years. Increasing inflow by such a factor is technically possible but economically, socially, and politically improbable since it would necessitate cutting consumptive irrigation withdrawals by nearly 50 per cent and reducing the irrigated area by around 45 per cent (assuming significant further improvements in irrigation water-use efficiency). However, a smaller Aral Sea with some economic and ecological importance could be preserved with considerably less inflow.

A number of schemes to accomplish this as well as to ameliorate environmental conditions around the sea have been proposed. The simplest and quickest approach would be to supplement the sea's water balance by channeling irrigation drainage water to it that is now lost to evaporation or accumulated in lakes. Perhaps 10–12 cubic kilometres of drainage water annually could be sent to the Aral by collectors paralleling the Amu and Syr Dar'ya which would also serve to keep this pesticide-, herbicide-, and defoliant-laden, saline flow out of the two rivers.[65] However, unless treated before release to the Aral, it would further degrade the sea's water quality. Preliminary work on a 1,500 kilometre collector along the Amu Dar'ya reportedly is underway.[66] Diverting irrigation drainage water to the sea will dry the two largest lakes supported from this source, Aydarkul' and Sarykamysh (figure 7.1). With an aggregate area over 5,000 square kilometres, they have developed considerable fishery and wildlife habitat significance which will be lost (although they are already suffering degradation from accumulating salts and toxics contained in irrigation drainage).[67]

Delivery of 12 cubic kilometres of irrigation drainage water plus 3–4 cubic kilometres of net ground water inflow would support a sea of only 18,000 square kilometres whose salinity would be so high (over 100 grams per litre) that its ecological and economic value would be near zero. Thus, additional measures are necessary to preserve the Aral in some useful form and to reduce the adverse impacts of its recession. The sea could be partitioned with dikes with low salinity preserved in those parts which received river inflow.[68] Several projects of this genre have been put forward by Soviet experts requiring minimum annual surface inflows from 8 to 30 cubic kilometres and maintaining an 'active', low salinity water body of 12,000 to 30,000 square kilometres.

Another approach is to focus on improving the ecologically and economically vital deltas of the Amu and Syr Dar'ya.[69] For the former, the plan is to construct a dike on the dried bottom facing the sea to create a shallow reservoir that would raise water levels in the delta, allowing partial restoration of its former ecological and economic value. The plan would require 12 cubic kilometres of water per year, mainly from irrigation drainage. The dried sea bed in front of the delta would be stabilised by plantings. The residual Aral Sea would stabilise near the 30 metre level (9 metres lower than in 1989). Estimated project cost is 406 million roubles. A similar plan could be implemented for the Syr Dar'ya delta, requiring some 7 cubic kilometres a year.

More water could also be provided to the Aral from the rivers of Western Siberia lying to the north.[70] A structural trough (the Turgai Gate) with a maximum elevation of 120 metres links the Arctic and Aral Sea drainage basins. Bringing Siberian water to the Aral, although ecologically disruptive, expensive, and a major engineering feat, is technically feasible. The diversion possibility was recognised even in Tsarist times, but the first serious schemes were formulated during the Soviet era, including the enormous Davydov project that would have flooded a huge portion of Western Siberia, caused tremendous ecological and economic harm, and cost the equivalent of several hundred billion roubles today.

The 1970s was a period of intensive development of water redistribution plans but with a greater focus on minimising their potential environmental impacts. By the end of the decade, detailed designs had been formulated for diversions in both the European and Siberian parts of the country. The first phase of a Siberian transfer was undergoing detailed engineering design in 1985 and was scheduled for implementation by the late 1980s or early 1990s.[71] The plan was to take 27 cubic kilometres annually from the Ob' and Irtysh rivers in Western Siberia. Water would be sent 2,500 kilometres southward through the Turgai Gate, into the Aral Sea basin and as far south as the Amu Dar'ya by a system of low dams, pumping stations, and a huge earth-lined 'Sibaral' (Siberia to Aral) canal (figure 7.1). Providing more water for irrigation was the scheme's main purpose (90 per cent was intended for this sector), but it would have helped stabilise the Aral as well, for example, by increasing irrigation return flows to the Amu and Syr Dar'ya rivers.

During the Gorbachev era, the fortunes of the European and Siberian schemes have waned. The concept of north–south water transfers has been bitterly attacked in the Russian popular media as ill-conceived, poorly planned, enormously expensive, and environmentally harmful.[72] In August 1986, a joint Party and government decree halted construction on the European project and planning for the Siberian undertaking.[73] However, research on the scientific, ecological, and economic problems associated with interbasin water transfers was to continue. A perception of excessive costs compared to expected benefits appears to have been the dominant factor in the cancellation of the projects.[74]

Nevertheless, the Siberian water transfer project may be revived, if, as is likely, regional water resources prove inadequate to meet future economic and social needs and preserve a viable Aral Sea. Central Asian water management experts are again arguing for water transfers from the north as the only means to save the region from a catastrophe.[75] The 19 January 1988 decree of the CPSU Central Committee and USSR Council of Ministers on improving water use in the country directed that scientific study of north–south water transfers continue.[76]

Political leaders in Central Asia (e.g. I. A. Karimov, President of the Uzbek Republic and First Secretary of the Uzbek Communist Party) were no doubt enormously disappointed but remained quiet for several years after the Siberian diversion project was halted. But by 1989, they were again stressing the dire nature of the water situation in Central Asia, calling for concrete help from Moscow to alleviate the problem, and raising the question whether the region can survive on its own water resources.[77] With the weakening of central authority and the declarations of sovereignty by the various union republics, Central Asian politicians have grown bolder and more adamant on this issue. On 23 June 1990, the presidents of the four Central Asian republics (Uzbekistan, Kirgizia, Tadzhikistan, and Turkmenia) and Kazakhstan signed a joint declaration on mutual problems and approaches to their solution.[78] The ecological catastrophe of the Aral Sea and adjacent area was cited as an acute problem that could not be solved by regional efforts alone. The leaders called on the national government to declare the Aral region one of national calamity and to provide real help. They also stated that it was necessary to return to the idea of water diversions from Siberia as one of the principal routes of saving the Aral and ensuring an adequate food supply for the region. In their view, diversions will decide the region's future.

This official, high level call for reconsideration of the Siberian water diversion project as one of the primary means to alleviate the water crisis in Central Asia is sure to further inflame this already bitterly contentious issue between the national government and the Russian republic on one side and the Central Asian republics on the other. This was evident during the visit of a UNEP (United Nations Environment Programme) working group on the Aral Sea problem to Central Asia in September 1990 when on a number of occasions local and regional officials went out of their way to stress the necessity of water transfers from Siberia, causing the members of the group from the Russian republic to publicly and adamantly disagree with them.[79]

A possible compromise (although probably not acceptable to the Russian republic) might be a considerably scaled-down version of the Siberian plan in which the diverted water would be intended specifically for the Aral and not for irrigation expansion. Water could be delivered to the northern part of the sea, shortening the route and reducing costs. The project could be defended as necessary to save the Aral – arguably a benefit outweighing any environmental harm caused in Western Siberia. Ten to fifteen cubic kilometres of fresh water from Siberia per year and implementation of local measures could probably preserve the Aral near its present size (40,000 square kilometres) and lower salinity to ecologically tolerable levels.

In any case, a national rescue effort has been mounted to save the Aral Sea.[80] Under pressure from the popular media, concerned scientists, and a 'Save the Aral Committee' (Komitet po spaseniyu Arala i Priaral'ya) organised by the Uzbek Writers' Union, the government issued a decree in September 1988 to implement a programme to ameliorate the Aral problem.[81] Based on recommendations from an expert commission, it directed that efforts be implemented between 1988 and 2005 to preserve the sea itself as well as improving the deteriorating ecological situation, drinking water supply, and human health conditions in areas adjacent to the Aral. Minimum guaranteed inflow to the sea is to steadily increase from 8.7 cubic kilometres in 1990 to 21 cubic kilometres in 2005 by means of water savings realised through efficiency improvements in irrigation and the delivery of irrigation drainage water. The Amu and Syr Dar'ya deltas are also to be preserved by schemes similar to those discussed above, and a plan developed to stabilise the dried sea bottom. The effort will be costly; one authoritative source has cited a figure of around 30 billion roubles for implementation.[82]

The decree is being taken seriously. Provisions have been made to guarantee financing, work on some elements of the plan (e.g. delivering more drainage water to the sea and improving drinking water supplies and medical services for the population) is underway, and oversight agencies have been established to ensure compliance of responsible organisations.[83] Nevertheless, vehement complaints are being heard from Central Asians that the programme is going slowly and poorly. Tulepbergen Kaipbergenov, the well-known Karakalpak writer, scathingly criticised efforts in his speech to the Congress of Peoples' Deputies on 2 June 1989 and stated that during the first five months of 1989 not one drop of river water had reached the Aral (inflow for all 1989 is estimated near 4 cubic kilometres).[84] Similar complaints were heard from Central Asians during the visit of the UNEP working group to the Aral Sea region in autumn 1990.[85]

Furthermore, even if the programme is fully implemented in a timely fashion, results may be disappointing. The stepped increase of minimum guaranteed annual inflow to the Aral to 21 cubic kilometres per year by 2005 will result in the sea's continued shrinkage, albeit at a decreasing rate into the next century. The sea would eventually stabilise at near 34 metres, 5 metres below the 1989 mark, and have an area around 27,000 square kilometres, 13,000 less than in 1989. Unless measures are taken to desalinate and detoxify irrigation drainage water delivered to the sea, its average salinity will not only be too high at near 70 grams per litre for valuable aquatic life, but the Aral will be seriously polluted as well. The economic and ecological usefulness of the sea would be practically nil.

International efforts are also underway to help with the Aral problem. In January 1990 the Soviet Government signed an agreement with UNEP for assistance in preparing an action plan for the rehabilitation of the sea.[86] The UNEP working group, composed of experts from the USSR and abroad, visited the Aral region in September 1990.[87] They will make several more fact-finding trips to Central Asia in 1991. The working group is to provide its recommendations to the Soviet government in January 1992. Two major international Conferences have also been held on the Aral problem: at Bloomington, Indiana in July 1990 and at Nukus, Karalkalpak ASSR, in October 1990.[88] Collaborative research on the Aral involving foreign scientists is also underway.[89]

Summary and conclusions

Soviet Central Asia has critical water management problems. Extensive development of irrigation has exhausted surface water resources, placing the future of irrigation, the region's economic foundation, in jeopardy. Population continues to grow rapidly, which requires an expanded water and food supply as well as increased employment opportunities. Large-scale emigration to labour-short regions in the Russian republic is one possible solution, but most ethnic Central Asians have no desire to relocate where climate, language, and culture are so different from their native lands.

Major efforts have been made since 1981, and accelerated under Gorbachev, to improve irrigation water use: reconstruction of antiquated irrigation systems, water application improvements, automation and computerisation, and shifting the crop mix toward lower water-consuming types. The programme has had some success but much remains to be done. It is planned to establish prices for irrigation water in the near future to promote its careful use. Critics of irrigation contend that the potential savings from efficiency improvements are enormous and more than enough to meet future regional needs. Water management specialists see only modest quantities of water being freed at great cost over a lengthy period.

Available water supplies could be increased somewhat by such measures as greater use of ground water, re-use of irrigation drainage, more regulation of river flow, and the use of water collected in natural basins and ephemeral streams. These measures should be developed but entail technical problems and economic and environmental costs. The economic structure of Central Asia could also be refocused: away from water intensive irrigation and toward low water-use industries (e.g. textiles, clothing, electronics). This holds promise but would require substantial increases in regional capital investment, retraining of the populace, and fundamental socio-cultural changes as society shifts from rural/agricultural to urban/industrial.[90] Such wholesale alteration of society would take decades. Furthermore, this strategy would not solve the problem of increasing local food production and, indeed, might lessen it as people moved from village to city.

The man-made drying up of the Aral Sea further complicates an already difficult water management situation. If preventive measures are not taken, it will shrink to several residual brine lakes in the next century. The sea's desiccation has caused progressively worsening environmental, economic, and social problems whose costs, although difficult to measure precisely, have certainly accumulated into the tens-of-billions of roubles already. A September 1988 governmental decree made amelioration of the Aral problem a national priority. But although a step in the right direction, the decree is proving difficult to implement and will be costly. Furthermore, it will only preserve a highly saline, much shrunken, and polluted water body. International efforts are also underway to provide assistance.

Until recently, hopes to resolve Central Asia's water problems rested on future massive water transfers from Western Siberia's rivers. This project was halted in 1986 pending further study and validation. Central Asian leaders, nevertheless, are again calling for its implementation, but the Russian republic, from where water would be diverted, continues to bitterly resist the idea. Without the Aral sea's difficulties there would be a reasonable probability for a successful regional strategy to cope with water problems associated with irrigation, population growth, and the need for economic development. But to deal adequately with both simultaneously without water importation seems extremely difficult. A compromise might be a scaled-down version of the project which would withdraw 10 to 15 cubic kilometres annually from Siberian rivers and send it directly into the northern part of the Aral sea. This, along with implementation of local measures, could preserve the sea near its current level and area while lowering salinity to ecologically tolerable levels – all without any significant cut-back in irrigation. It could be argued that saving the Aral outweighs the harm to Western Siberia (although inhabitants of that region, no doubt, would take exception). The Soviet government and Russian republic could, as a condition of the 'deal', stipulate that no Siberian water be used for irrigation, encouraging Central Asian water interests to be more efficient, since expansion of irrigation and other water uses would be possible only from water freed by this means.

There are no simple, cheap, or easy answers to the Central Asian water situation. Irrigation expansion based on local water resources, in spite of efficiency measures, will stop in the 1990s. This alone may cause severe social and economic disruption and force large-scale emigration. To provide the inflow the Aral requires to remain ecologically viable, irrigation would have to be substantially reduced, inevitably inducing these consequences. Long-lasting regional enmity toward the central government could result, perhaps leading the Central Asian republics and Kazakhstan to leave the USSR. On the other hand, substantial aid from Moscow in improving the situation could be a major factor inducing the region to remain with the Union.

Notes

1. *Atlas SSSR* (Moscow: GUGK, 1983), p. 3.
2. G. V. Voropayev, B. G. Blagoverov, and G. Kh. Ismayilov, *Ekonomiko-geograficheskiye aspekty formirovaniya territorial'nykh edinits v vodnom khozyaistve strany* (Moscow: Nauka, 1987), pp. 212–15.
3. Ibid.
4. Estimate based on extrapolation of data in *Narodnoye khozyaistvo SSSR, 1980–1988*; *Narodnoye Khozyaistvo Kazakhstana*; 1984, 1986, 1988; *Narodnoye Khozyaistvo Kirgizskoi SSR*, 1988.
5. N. F. Vasilev, 'Land reclamation: in the front line of perestroika', *Gidrotekhnika i melioratsiya*, no. 11 (1987), pp. 2–10.
6. V. A. Dukhovnyi and R. M. Razakov, 'Aral: Looking the truth in the eye', *Melioratsiya i vodnoye khozyaistvo*, no. 9 (1988), pp. 27–32.
7. *Pravda*, 27 October 1984, pp. 1–2.
8. See the sources cited in note 4 above.
9. 'Concerning the first priority measures for improving the use of water resources in the country', *Vodnyye resursy*, no. 6 (1988), pp. 5–20.

10. N. F. Glazovsky, 'A conception of the way out of the "Aral Crisis"', *Izvestiya Akademii Nauk, seriya geograficheskaya*, no. 4 (1990), pp. 28–41; and A. K. Kiyatkin, 'The second session of the Aral movement', *Melioratsiya i vodnoye khozyaistvo*, no. 1 (1990), pp. 8–10.
11. Philip P. Micklin, *The Water Management Crisis in Soviet Central Asia*, final report to the National Council for Soviet and East European Research, contract No. 802–09, 15 February 1989, pp. 17–27; Dukhovnyi and Razakov, 'Aral'.
12. *Pravda*, 14 April 1988, p. 3.
13. Voropayev, Blagoverov and Ismayilov, *Ekonomiko-geograficheskiye aspekty*, pp. 212–15.
14. V. A. Dukhovnyi, I. S. Avakyan, and V. V. Mikhailov, 'Reclamation, water management and social-economic problems of Central Asia', *Melioratsiya i vodnoye khozyaistvo*, no. 9 (1989), pp. 3–6.
15. See the sources cited in note 4 above.
16. Dukhovnyi and Razakov, 'Aral', pp. 27–32.
17. Ibid.
18. Ye. Nesterov, 'Aral: time for action', *Melioratsiya i vodnoye khozyaistvo*, no. 11 (1988), pp. 1–4.
19. Polad-Polad Zade, First Deputy Minister of Minvodstroy, interview, Moscow, 14 September 1989.
20. Micklin, *Water Management Crisis*, pp. 27–30.
21. B. G. Shtepa, *Tekhnicheskii progress v melioratsii* (Moscow: Kolos, 1983).
22. Ye. P. Gusenkov, V. A. Dukhovnyi, and A. I. Tuchin, 'Main trends in the rational use of water resources in Central Asia and Kazakhstan', a paper presented to the Sixth IWRA World Congress on Water Resources, Ottawa, Canada, 29 May–3 June 1988.
23. V. Ya. Lashchenov, 'Problems of the interrepublic distribution of the Syr Dar'ya's waters', *Melioratsiya i vodnoye khozyaistvo*, no. 1 (1990), pp. 3–5.
24. Ibid.
25. M. Vagapov and D. Azimov, 'Reconstruction of irrigation systems – the foremost task', *Melioratsiya i vodnoye khozyaistvo*, no. 3 (1988), pp. 16–18.
26. Lee H. Raymond and Kelly V. Rezin, 'Evapotranspiration estimates using remote-sensing data, Parker and Palo Verde Valleys, Arizona and California', *U.S. Geological Survey Water Supply Paper 2334* (Washington, 1989), 18 pp.
27. V. A. Dukhovnyi, 'Economize on irrigation water!' *Gidroteckhnika i melioratsiya*, no. 5 (1985), pp. 40–3.
28. Voropayev, Blagoverov and Ismayilov, *Ekonomichesko-geograficheskiye aspekty*, pp. 182–3.
29. I. M. Chernenko, 'Problems of managing the water-salt regime of the Aral Sea', *Problemy osvoyeniya pustyn'*, no. 1 (1986), pp. 3–11.
30. G. V. Voropayev, G. Kh. Ismayilov, and A. A. Bostandzhoglo, 'Intensification of the use of water-land resources of the river basins of the Aral Sea with consideration of the achievements of scientific-technical Progress', *Obshchestvenniye nauki v Uzbekistane*, no. 3 (1988), pp. 5–14.
31. N. S. Bystritskaya, 'Concerning the transfer of operational water management organizations to economic accounting', *Melioratsiya i vodnoye*

khozyaistvo, no. 9 (1988), pp. 7–8; and N. S. Bystritskaya and R. A. Krivov, 'Concerning payment for irrigation water', *Melioratsiya i vodnoye khozyaistvo*, no. 6 (1988), pp. 8–9.
32. *Pravda*, 26 September 1988, p. 4.
33. A. Kes', 'Reasons for the level changes of the Aral during the Holocene', *Izvestiya Akademii Nauk SSSR, seriya geograficheskaya*, no. 4 (1978), pp. 8–16; and D. B. Oreshkin, *Aral'skaya katastrofa, Nauka o zemle*, no. 2 (Moscow: Znaniye, 1990), pp. 3–17.
34. Philip P. Micklin, 'Desiccation of the Aral Sea: a water management disaster in the Soviet Union', *Science*, 241, no. 4870 (2 September 1988), pp. 1170–6.
35. D. Ya. Ratkovich, 'Concerning the problem of ensuring the country with water in light of the requirements of nature protection', *Vodnyye resursy*, no. 5 (1989), pp. 5–16; R. M. Razakov, 'Ecological measures in the Aral region: investigations and an action program', *Melioratsiya i vodnoye khozyaistvo*, no. 1 (1990), pp. 6–8; and Z. D. Tkacheva, 'Dynamics of the Aral Sea from 1957 to 1989 with a forecast to 2000', 1:1,000,000 map (Moscow: GUGK, 1990).
36. *Pravda Vostoka*, 10 September 1987, p. 3 and 2 June 1989, p. 1.
37. Micklin, 'Desiccation of the Aral Sea', pp. 1170–6.
38. Ibid.; Micklin, *Water Management Crisis*, pp. 56–7; and I. A. Shiklomanov, *Antropogenye izmeneniya vodnosti rek* (Leningrad: Gidrometeoizdat, 1979), p. 227.
39. Micklin, *Water Management Crisis*, pp. 57–8.
40. B. T. Krista, 'The problem of the Aral Sea and the Karakum Canal', *Problemy osvoyeniya pustyn'*, no. 5 (1989), pp. 10–17.
41. Micklin, *Water Management Crisis*, pp. 58–9; Institut geografii, *Problemy Aral'skogo morya*, ed. S. Yu. Geller (Moscow: Nauka, 1989), pp. 5–25.
42. Micklin, 'Desiccation of the Aral Sea', pp. 1170–6.
43. Ibid.
44. Gosudarstvennyi komitet SSSR po gidrometeorologii and Gosudarstvennyi okeanograficheskii institut, *Gidrometeorologiya i gidrokhimiya morei SSSR*, tom VII: *Aral'skoye more* (Leningrad: Gidrometeoizdat, 1990), pp. 26–28.
45. Micklin, 'Desiccation of the Aral Sea'.
46. Micklin, *Water Management Crisis*, p. 61.
47. A. A. Tursunov, 'The Aral Sea and the ecological situation in Central Asia and Kazakhstan', *Gidrotekhnicheskoye stroitel'stvo*, no. 6 (1989), pp. 15–19.
48. Micklin, 'Desiccation of the Aral Sea', pp. 1170–6.
49. Razakov, 'Ecological measures', pp. 6–8.
50. M. Ye. Bel'gibayev, 'The dust/salt meter – an apparatus for capturing dust and salt in air streams', *Problemy osvoyeniya pustyn'*, no. 1 (1984), pp. 72–4; and Gennadii Shalayev, *Sobesednik*, no. 32 (August 1990), pp. 8–9.
51. Tursunov, 'The Aral Sea', pp. 15–19.
52. N. V. Aladin and S. V. Kotov, 'The natural condition of the Aral Sea ecosystem and its anthropogenically induced change', pp. 4–25 in Trudy Zoologicheskogo Instituta, AN SSSR, *Gidrobiologicheskiye problemy Aral'skogo morya*, tom 99 (Leningrad, 1989).

53. Micklin, 'Desiccation of the Aral Sea', pp. 1170–6.
54. Ibid.
55. Ibid.
56. Razakov, 'Ecological measures', pp. 6–8.
57. *Pravda Vostoka*, 3 June 1989, p. 2.
58. Alimbetova, Khadisha, deputy director of the Muynak hospital, interview, Muynak, 7 September 1989.
59. Micklin, *Water Management Crisis*, p. 66.
60. Alimbetova interview.
61. Philip P. Micklin, 'Touring the Aral: visit to an ecologic disaster zone', *Soviet Geography*, XXXII, no. 2 (1990), in press.
62. Micklin, *Water Management Crisis*, p. 66.
63. Dukhovny and Razakov, 'Aral', pp. 27–32.
64. V. Kovalev, 'Irrigation and the Aral', *Zvezda Vostoka*, no. 12 (1986), pp. 3–15.
65. Micklin, *Water Management Crisis*, p. 68.
66. *Izvestiya*, 27 September 1987, p. 2, and *Pravda Vostoka*, 7 July 1989, p. 2.
67. A. K. Kiyatkin and M. V. Sanin, 'Sarykamysh Lake – the largest storage basin for collector-drainage water', *Melioratsiya i vodnoye khozyaistvo*, no. 1 (1989), pp. 20–4.
68. Micklin, *Water Management Crisis*, pp. 69–71.
69. Ibid., pp. 70–1.
70. Ibid., pp. 71–81.
71. Philip P. Micklin, 'The status of the Soviet Union's north–south water transfer projects before their abandonment in 1985–86', *Soviet Geography*, XXVII, no. 5 (1986), pp. 287–329.
72. P. Micklin, 'The fate of "Sibaral": Soviet water politics in the Gorbachev era', *Central Asian Survey*, no. 2 (1987), pp. 67–88; and Philip P. Micklin and Andrew R. Bond, 'Reflections on environmentalism and the river diversion projects', *Soviet Economy*, 4, no. 3 (1988), pp. 253–74.
73. *Pravda*, 20 August 1986, p. 1; and Ziyadullayev, S. K., 'Utilization of the land and water resources of Central Asia and southern Kazakhstan', *Problemy osvoyeniya pustyn*', no. 2 (1990), pp. 3–7.
74. Micklin, 'The fate of "Sibaral"', pp. 67–88.
75. Micklin, *Water Management Crisis*, pp. 79–81, and Ziyadullah, 'Utilization', pp. 3–7.
76. 'Concerning the first priority measures', pp. 5–20.
77. *Pravda Vostoka*, 23 September 1989, pp. 1–2, and 1 December 1989, p. 2.
78. *Pravda Vostoka*, 24 June 1990, p. 1.
79. Micklin, 'Touring the Aral'.
80. Micklin, *Water Management Crisis*, pp. 81–90.
81. *Pravda*, 30 September 1988, pp. 1–2.
82. Kungrad Doshumbayev, Deputy Director of Aralvodstroy [Aral water management construction], interview, Nukus, 8 September 1989.
83. *Pravda Vostoka*, 14 May 1989, p. 2.
84. *Pravda Vostoka*, 3 June 1989, p. 2.
85. Micklin, 'Touring the Aral'.
86. Centre for International Projects, State Committee for Environmental Pro-

tection. *Assistance for the Preparation of an Action Plan for the Rehabilitation of the Aral Sea*, USSR/UNEP Project (review of the problem and brief information on the project) (Moscow, 1990).
87. Micklin, 'Touring the Aral'.
88. 'The Aral Sea crisis: environmental issues in Central Asia', an international conference sponsored by the Research Institute for Inner Asian Studies and the School of Public and Environmental Affairs, Indiana University, Bloomington, Indiana, 14–19 July 1990; and 'The Aral crisis; causes, consequences, and ways of solution', an International Symposium sponsored by the Special Research and Coordination Centre 'Aral', Nukus, USSR, 2–5 October 1990.
89. Micklin, 'Touring the Aral'.
90. Dukhovny, Avakyan and Mikhaylov, 'Reclamation, water management and social-economic problems', pp. 3–6.

8 Perestroika: how it affects Soviet participation in environmental cooperation

Elena Nikitina

Perestroika, currently underway in the USSR with its political reform, transition to market economy, restructuring in economic management and other revolutionary innovations, has changed general approaches to the environment. It has led to progress both in the transformation of national nature conservation activities and the development of a new international environmental policy.

For a rather long period there were serious divergences between the declared principles of nature and socialist society's interactions with the real methods of environmental management, and these resulted in serious nature degradation. The economic damage from industrial pollution of the environment in the USSR, according to estimates of the Soviet Academy of Sciences, comprises annually about 50–70 billion roubles or 0–11 per cent of national income.

But now a new environmental policy is evolving. Its major points envisage: changes in economic activities to provide environmentally safe economic growth; the rationalisation of consumption to combine greater economic well-being with preservation of the environment and its resources; the incorporation of common values of mankind into environmental policy; the improvement of environmental legislation; and the development of environmental education at all levels.[1] Dramatic changes in national environment protection management have the closest connections with new international environmental approaches.

In the late 1980s the Soviet Union began to advocate a strategy of international cooperation on environmental protection which envisages joint efforts for the protection of values common to all mankind, since the global destruction of nature demands multilateral solutions to provide for the environmentally sustainable development of humanity. Conceptually, this strategy is based on the principles of International Environmental Security (IES) which foresees a higher level of environmental interaction and integration of states. At this level, protection, preservation and improvement in the quality of the environment as well as sustainable economic growth are secured, and elaboration by mutual efforts of measures counteracting global and regional ecological threats – corresponding to national and collective interests of the participants – are achieved. It can be assured by implementation of an agreed set of concrete international measures as well as successive national efforts.

Implementation of the IES concept envisages a wider use and compliance with the existing institutional arrangements of environmental cooperation, as well as formation of new effective international mechanisms and procedures, the adoption of basic rules and principles for the conduct of states, the introduction of international control for adherence to these rules, the outlining of responsibility/liability principles, and the establishment of prior directions of interrelation of states.

The first outlines of this concept which is in the initial stage of its development were formulated by the Soviet Union together with certain countries of Eastern Europe in mid-1988.[2] Since then it has been developed further and been under discussion at large international forums with some advocates as well as opponents. The problems of its implementation were considered in the framework of the United Nations – during the 42nd and 43rd sessions of the General Assembly and in the UN Environmental Programme.

In recent years the Soviet Union has proposed certain international mechanisms to implement an IES régime. They include, for instance, the creation of new institutional structures and mechanisms as an international centre of assistance in cases of environmental emergencies, international régimes of environmental monitoring, governmental reporting on national nature preservation activities, and multilateral exchange of environment-saving technologies. The USSR has stressed the need for adopting during the forthcoming 1992 UN Conference on Environment and Development an International Code of Environmental Ethics that would combine the rights and obligations of states towards nature and each other on this issue. The elaboration and implementation in practice of principles of international responsibility/liability for environmental damage are supposed to be one of the major items on the international agenda in the near future.

A special part of the Soviet state's environmental programme for 1991–5 and up to 2005, finally issued by Goskompriroda at the end of 1990, is dedicated to the prospects of Soviet participation in international environmental cooperation. Although rather declarative in its character, it indicates, however, the concrete spheres of prior Soviet interest in the framework of multilateral efforts. They include *inter alia*, the formation and development of international régimes on biodiversity, global warming and ozone layer protection, transboundary environmental issues (including air, water pollution and consequences of industrial accidents with transboundary effects), protection of the unique nature of the Arctic and the Antarctic, and the Black Sea environment.[3]

These are not the only elements comprising the essence of formation of Soviet international environmental policy. Certain positive and dramatic results have been already achieved in addressing *global environmental changes*, particularly regarding the problems of ozone layer depletion, global climate change and transboundary air pollution. They are in line with the latest developments towards international régimes which, in comparison with many previous international efforts, represent a higher stage of environmental interaction between states. They envisage, for instance, international environmental management with specific mutual obligations of states with defined time and volume scales. These could be considered as a rather progressive and effective approach to international regulations of environmental activities and could serve to some extent as a model for future developments in nature preservation cooperation.

Serious practical implications of the new international environmental approaches can be observed at a national level: an interesting and topical subject for detailed scientific investigation. The formation of these mechanisms foresees deep interactions of international response measures with national economies entailing even their restructuring and the creation of additional institutional mechanisms and surplus financial investments. They presuppose also the elaboration of legal norms and procedures as, for example, states' responsibility/liability norms with further national compensation for environmental damage and even pollution taxation mechanisms. These serious problems are now under consideration and are accompanied by opposing attitudes among different groups of states. Regrettably the approach to this issue both of the USSR and other major developed nations has not yet been finally determined.

During recent years the Soviet Union has actively contributed to the efforts to establish balanced international policies on global environmental change issues that would serve as a key item in international environmental cooperation for the near future. Let us illustrate current Soviet involvement in international efforts by addressing three major topics of the global environmental change agenda, examining their particular influence on the national economy and their interrelation with national environmental policies and perestroika. They are: depletion of the ozone layer, global climate change, and transboundary air pollution.

Three major issues

Depletion of the ozone layer

The Soviet Union accounts for about 12 per cent of world production of ozone-depleting substances (EEC 40 per cent, USA 35 per cent, Japan 10 per cent, developing countries 2 per cent) and has participated in the elaboration and implementation of international policies regarding this problem. It is a party to the 1985 Vienna Convention for the Protection of the Ozone Layer and its 1987 Montreal Protocol on substances that deplete the ozone layer, requiring the reduction by 1998 of the production and use of CFCs by 50 per cent of the 1986 levels. As it has become apparent that this 50 per cent target will not be adequate to stop the ozone layer depletion, the Soviets seconded the motion by some developed countries for stronger obligations than those envisaged by the Montreal Protocol – that is, the complete cessation of CFCs and halon production by the end of the century. The USSR, among other states, approved in mid-1990 in London a protocol for a total ban on the production of major ozone-depleting chemicals by the year 2000. Besides this, the USSR is involved in the fourteen-member executive committee which is to be created under this new arrangement to administer a new international fund to help poor countries move to technologies free of CFCs.[4]

Faced with a variety of possible strategies to provide its obligations according to the ozone layer protection international régime (changes in consumption structures, recycling, a shift to production of alternative non-depleting substances) the Soviet Union is to pursue the most radical measures. The problem of converting enterprises to the production of non-active ozone substances is one item on its agenda. This process began in 1988–1989 when three aerosol enterprises transferred their production to non-active propellants. A special national mechanism – an interdepartmental commission of the Council of Ministers – was created to effect the transformation of certain branches of the chemical industry in compliance with international obligations, and the 1989–1995 programme of ozonosphere research and monitoring was adopted.

Global climate change

Recent years have been marked by the mutual efforts of states to outline a common strategy for humanity on global climate change primarily by limiting industrial carbon dioxide emissions. In the mid 1980s the Soviet Union ranked second in the world after the USA in industrial carbon dioxide discharges and accounted annually for about 23 per cent of the world total of 958 million tons (USA 23 per cent, Western Europe 15 per cent, China 10 per cent, developing countries 2 per cent).

The development of a common strategy of cooperation including scientific substantiation of the problem and international mechanisms to tackle global warming is one of the pressing items of the current international environmental agenda. One of the prior tasks here is to formulate the international obligations of states to reduce the emissions of greenhouse gases into the atmosphere, which would entail agreement on regulatory norms and standards with subsequent adoption of national actions.

The Soviet policy on this issue, involving both participation in international efforts and developments on a national level, is now being evolved and a special national long-term programme on global warming issues is being elaborated. Its preliminary outlines embrace a wide range of actions including: scientific research and climate modelling; evaluation of the global climate change impact on national economies, population and environment; the search for adaptation measures; the planning of major industries' development with emission reductions; participation in the creation and implementation of an international framework convention with a series of protocols on global climate change.[5]

Though the tasks are rather diversified, we believe that in the nearest future comparatively greater attention will be paid at national level to the formulation of concrete measures which will counteract the negative impact of global warming as well as slow down air pollution and the consequent changes in the atmosphere. There is certainly enormous potential in the Soviet Union for reducing carbon dioxide emissions, for instance by increasing energy efficiency with restructured energy production and consumption plus conservation efforts. On a per capita basis, primary energy and electricity use in the Soviet industrial sector is now about one-quarter higher than in the United States. Current perestroika and modernisation are supposed to make the industrial sector far less materials and energy intensive, as would also, of course, improved product quality and maintenance. According to some specialists, the USSR should encourage the development and use of new technologies through a combination of energy-pricing mechanisms and requirements, such as high standards for efficiency. During recent years the national energy balance structure has shifted from coal to natural gas (which produces the greatest amount of energy per unit of CO_2 emissions of any fossil fuel) by a proportion of 24 per cent in 1980 to 41 per cent currently.[6] The development of alternative sources of energy (geothermal, solar, tidal, wind power), now under way in the Soviet Union, would account for 5 per cent of energy production by the beginning of the twenty-first century and could be regarded as a definite part of the solution of the global warming problem. Until recently more dependence was being placed on atomic power (now providing about 12 per cent of electricity production), but this has changed with the Chernobyl tragedy and the growing opposition from the Green movement. With the implementation of some of these measures, the Soviet Union's CO_2 emissions have been reduced by 3 per cent in recent years – from 15.4 million tons in 1985 to 14.9 million tons in 1989.

Transboundary air pollution

A significant part of the Soviet environmental policy deals with the problem of acidification as a result of long-range transboundary air pollution. The Soviet Union participated in the creation of an international regulation mechanism in the framework of the 1979 Convention on Long-Range Transboundary Air Pollution which, with its three protocols, represents one of the first successful examples of international environmental cooperation. On the regional scale, since 1980 SO_2 emissions and their transboundary fluxes in EEC countries have declined by about 20 per cent and this downward trend is expected to continue.

The annual deposition of sulphur and nitrogen oxides on Soviet territory runs today at about 23 million tons, while the pollution and acidification of the western regions of the European USSR are higher than the country's average. The emissions of sulphur dioxide are comparatively high (about 30 per cent less than in the USA) and equal the total of Germany, the UK and Czechoslovakia combined – Europe's major exporters of acid rains. In 1989 Soviet industry discharged 17 million tons of sulphur dioxide and 5 million tons of nitrogen oxides into the atmosphere. Yet, despite all this, the USSR appears to be a net-importer of air pollutants. Due to the prevailing western air currents, the amount of pollutants imported from Western Europe, according to different estimates, is 5–6 times higher than that exported. A negative fluxes' balance occurs only in the north-western parts of the country: some enterprises of the Kola Peninsula are supplying the neighbouring states of north-west Europe with acid rain.

The compliance with international obligations within the framework of the 1979 Convention has serious economic consequences for the Soviet Union. The 30 per cent reduction, envisaged by the SO_2 Protocol, amounts to 5.2 million tons and is several times higher than the targets for each European country. According to the State Environmental Programme, installation of nitrogen oxides and sulphur dioxide industrial purification facilities by 1993 are estimated to cost 103 million and 236.4 million roubles respectively.

In the USSR in general during recent years certain reductions in total air pollution have been achieved, including reduction of sulphur dioxide and nitrogen oxides industrial emissions. The energy industry, one of the primary sources of air pollution, accounts for 65 per cent of sulphur dioxide and nitrogen oxides emissions and non-ferrous metallurgy for 25 per cent of sulphur oxides emissions.[7] The cut in sulphur dioxide emissions was obtained primarily from the increasing efficiency and structural changes in the energy sector, including the installation of purification facilities and increase in the use of natural gas. The emissions reduction target imposed for this sector for the late 80s was surpassed by 4 per cent.[8] But in non-ferrous industry the discharges were reduced by only 3 per cent in 1988 compared with the previous year. The current situation regarding emissions of nitrogen oxides is worsening, as they have increased over recent years (although still 2.5 times lower than in the USA), and certainly it needs the introduction of urgent regulatory measures as the present ones are inadequate to solve the problem.

However, according to the projected balance of sulphur dioxide discharges from different branches of industry, it is thought that the USSR will commit itself to the obligations envisaged by the SO_2 Protocol and by 1993 reduce its emissions and transboundary fluxes in general by 30 per cent (compared with the 1980 level) and by 1995 make 50 per cent reductions in the regions bordering with Finland. It is supposed that nitrogen oxides emissions would be stabilised by 1994 at the 1987 level, with their further reduction in compliance with the NOx Protocol.

One important question in this field is how to solve the problem of transboundary pollution from the Kola Peninsula's non-ferrous metallurgical enterprises 'Severonikel' and 'Pechenganikel'. These use ore with high sulphur content and have inadequate levels of purification facilities and sulphur utilisation from the metallurgical gases which emit annually about 500,000 tons of SO_2.[9] It is thought that by the mid-90s the discharges of sulphur oxide and correspondingly their transboundary fluxes will be reduced due to technological reconstruction of the metallurgical process provided by northern neighbours, suffering from transboundary pollution from the Kola Peninsula. Owing to the installation of modern facilities by the Finnish metallurgical company Outokumpu, the 'Pechenganikel' smelter's sulphur dioxide emissions will be reduced to one-fifteenth by 1994.[10] Financing will be sought through a joint Finnish–Norwegian–Swedish credit of 3 billion Finnish marks which will be repaid by the export deliveries of nickel from this reconstructed enterprise.

So the major problems in dealing with transboundary air pollution in the Soviet Union involve restructuring several branches of industry. They include not only the introduction of purification facilities and new environment-saving technologies to reduce harmful emissions into the atmosphere, primarily in the non-ferrous and energy sector, but also measures aimed at increasing energy efficiency. These measures envisage the use of alternative sources of energy, changes in the energy balance by moving to a wider use of natural gas, technological advances in burning fuel with a lower sulphur content and technological improvements in non-ferrous industries.

Changes in national environmental mechanisms

The compliance of Soviet international obligations with international environmental régimes is also determined by domestic factors. It is, too, closely connected – and should be combined with – measures to settle the increased environmental problems inside the country and with the elaboration of new national environmental management mechanisms with their economic, legal and organisational elements.

Indeed, the deterioration of the environment in the USSR has acquired threatening dimensions in certain regions, and according to some estimates it may even deteriorate further. About 16 per cent of Soviet territory can be considered as ecologically insecure,[11] while the Aral and Chernobyl zones are regions of ecological catastrophe. The level of harmful emissions into the air can be compared with that of the USA, although their indices per unit of national income are several times higher in the USSR. The concentrations of some pollutants in the air of major industrial regions and cities in the Soviet Union are dangerous for human health. According to official statistics, the level of air pollution in 103 cities during certain periods of 1990 exceeded the allowable concentrations more than 10-fold.[12] Due to the inadequate technological provisions of environmental protection activities, not much more than half of all industrial enterprises are equipped with air purification facilities. Due to the ineffectiveness of water purification installations or even their non-existence, only about one-quarter of discharged sewage is purified to the envisaged norms. The consumption of raw materials, energy and water per unit of production is several times higher than in many other developed countries. Nowadays the share of environmental investments in total capital spending is lower than in some Western countries (1.7 per cent in comparison with 4–5 per cent in the USA) and the major problem here is that only about half the financial resources apportioned annually to environmental purposes are used.

This alarming situation is a result of interactions of a wide range of economic, political and legal factors, one of the most important of which is misguided economic management practices. Wasteful economic mechanisms are aimed at maximising production, not preserving the environment. State enterprises and ministries are not made responsible for environmental damage. For instance, current pricing mechanisms of natural resources provide no incentive for their rational use or for the consumption of resources with less harmful impact on nature. Until recently producers had no economic obligations or interest in either conservation or purification. The result was the excessive use of natural resources and high levels of pollution. Formerly the situation was aggravated by the distortion in the legislative and executive power structures and the lack of real authority held by the local soviets, which were supposed to possess the functions of environmental preservation, in reality suppressed by the interests and powers of numerous ministries. Environmental legislation was also inadequate.

The complex of environment-saving measures for the amelioration of the Soviet Union's poor environmental situation and its compliance with international environmental obligations are included in the new system of environmental management being introduced in the USSR and outlined in the state environmental programme. The major objective is to incorporate the economic mechanisms of environmental management which regulate diversified interactions between man and biosphere and provide for environmentally sustainable development. Only their functioning in combination with progressive environmental legislation can lead to positive results and make those who damage the environment responsible and liable. This sort of environmental perestroika is now underway and some developments are already noticeable. They include: payments for natural resources and pollution; responsibility and liability of enterprises and ministries for pollution; transfer of environmental management from branches of industry to territorial administrations (all-union, republic, region, local); independence of local soviets from ministries as regards regional environmental funds; rationalisation of the financial mechanisms of environmental management; independent environmental impact assessment.

The compliance of Soviet international obligations within the framework of international environmental régimes is closely connected with the major trends in national environmental management. Further progress is affected by the current revolutionary developments in environmental policy, the 'greening' of the authorities and public opinion. Perestroika with its economic reform now in progress opens up great possibilities here, and its success determines to a great extent the prospects of Soviet participation in the international division of labour on environmental issues.

Notes

1. *Pravda*, 20 January 1990.
2. *Ekonomicheskaya gazeta*, July 1988, p. 1.
3. *Ekonomika i Zhizn'*, no. 41 (October 1990), p. 7.
4. *New York Times*, 30 June 1990, p. 2.
5. *Ekonomika i Zhizn'*, no. 41 (October 1990), p. 7.
6. *Narodnoye Khozyaistvo SSSR v 1989* (Moscow, Finansy i Statistika, 1990), p. 281.
7. *Doklad o sostoyanii prirodnoi sredy v SSSR v 1988* (Moscow, Goskompriroda, 1989), p. 54, 58, 61.
8. *Vestnik statistiki*, Moscow, 1988, no. 11, p. 56.
9. *Izvestiya*, 10 October 1990.
10. *Pravda*, 30 August 1990.
11. *Rodina*, no. 4 (1990), p. 69.
12. *Ekonomika i Zhizn'*, no. 5 (January 1991), p. 11.

9 US–Soviet cooperation for environmental protection: how successful are the bilateral agreements?

Kathleen E. Braden

Introduction

With the opening up of information sources from the USSR in the era of glasnost and perestroika, new statistics indicate that the Soviet Union's record has been disappointing in the realm of environmental disruption, confirming observations noted many years ago by Western scholars.[1] Is the dismal record of the Soviet bloc with respect to the environment a by-product of planned economic systems? Will the attempt to move the Soviet economy into more use of markets remove the roadblocks to environmental improvements?

Recent revelations about the state of the Soviet environment are emerging at a time when the United States and USSR are both developing new understandings of the global nature of environmental disruption. Perhaps a more useful question is: why do two countries with opposing economic systems arrive at similarly poor records of environmental safeguards? On the US side, gains in the realm of cleaner air and water must still be set against problems of acid rain, destruction of old growth forests, oil spills, toxic waste disposal, and other continuing dilemmas. Is a Western-style mixed market economy a useful model then for the Soviet Union?

In considering the convergence of US and Soviet societies in terms of environmental damage, an examination of a common touch-point between the two systems might prove helpful in revealing the potential for transfer of knowledge and technology to effect amelioration. One such 'laboratory' experience is the US–USSR Joint Agreement on the Environment. Activities covered under this agreement provide a seventeen-year history of cooperation to tackle jointly issues of both basic and applied science on behalf of an improved environment.

The record of accomplishments under the Joint Agreement is tracked and analysed here to ascertain whether the cooperation has yielded tangible benefits to one or both societies. If indeed perestroika in part seeks to impose more Western-style management on the USSR, then one would expect the agreement projects to provide US models for Soviet application to environmental improvement.

The following questions were considered:
1. What criteria for success are suggested in measuring the results of the agreement?
2. What has been the trend line for number of projects, number of activities, and the share for each of the ten Areas of the agreement? Have there been periods during which political considerations interrupted the work under the agreement?
3. What types of activities tend to be the most common, and how are they allocated among the various Areas?
4. Are the activities evenly distributed across regions of the USSR? How might a devolution of the USSR impact the work of the Joint Agreement?
5. Which jurisdictions seem to be most commonly responsible for overseeing the various projects?
6. How reciprocal have the activities been? Is equality maintained in terms of personnel and technology exchanges?

Answers to these queries may provide a more systematic evaluation framework for the success of the Agreement than anecdotal discussion of individual projects. In addition, clarity about success in different areas may suggest a sharpened focus for future work, particularly if the USSR seeks more transfer of environmental management philosophy and style from the West in the future.

Background to the US–USSR agreement on cooperation in the field of environmental protection

The United States and USSR have initiated many government-to-government agreements in the post-World War II era, enhanced by the 1958 General Exchanges Agreement.[2] Accords on science and technology were promoted with the International Geophysical Year, also in 1958, and with the Nixon–Brezhnev agreements of 1972. Environmental issues were inherent in many of these early cooperative programmes, such as agreements on study of the world's oceans, agricultural development, and public health. In addition, the US–USSR Cooperative Science Programme began in 1972, with exchanges between individual scientists of the two countries.[3]

Among the bilateral agreements to originate in the 1972 Moscow summit meeting was the US–USSR Agreement on Cooperation in the Field of Environmental Protection (referred to here as 'the

Environmental Agreement'). It has proved to have one of the best records of longevity of any of the 1972 agreements. Consisting of seven articles, the Environmental Agreement was signed by President Nixon and Chairman Podgorny on 23 May 1972, and in Article 5 sets up a Joint Committee to meet once a year, alternating between Washington and Moscow, to approve programmes and designate responsible organisations.[4]

Each year the Joint Committee issues a memorandum outlining planned projects, and a follow-up implementation report on the results of the previous meeting's projects. Article 2 established eleven major areas of cooperation, a list of areas which has been kept intact with new areas suggested only in 1985.[5] The most recent memorandum for the Twelfth Meeting of the Committee, signed in January of 1990, proposes a new Area XII, 'Information, education, and training in the field of environmental protection'.[6] The appendix shows projects under the eleven areas during four different years of the agreement to allow the reader to compare project titles over time.

Traditionally, the two agencies responsible for overseeing the Environmental Agreement have been the Environmental Protection Agency (EPA) in the United States and the USSR State Committee for Hydrometeorology and Environmental Control (Hydromet), but in 1988 the new State Committee on Protection of the Environment (Goskompriroda) took over the lead role on the Soviet side. In 1990, those chairing the committee were William Reilly of the EPA and Valentin Sokolovsky of Goskompriroda, and the actual work of overseeing the projects on the American side was carried out via the EPA Office of International Activities.

While work on many projects has proceeded remarkably uninterruptedly since 1972, the Environmental Agreement itself has not been immune to international geopolitical events. The agreement was extended for a five-year period in 1977, but when the USSR invaded Afghanistan, the committee itself fell on hard times as the US reduced many fields of cooperation. The advent of the Reagan administration did little initially to restore the activities. EPA administrator Ann Gorsuch did not schedule a meeting of the committee, and from January 1980 until June 1984 the Bilateral Agreement experienced a hiatus. Thus, the Eighth Meeting took place in late 1979, but the Ninth did not issue a memorandum until 1986. The memorandum at the time of this writing is therefore the Twelfth, despite the fact that eighteen years have elapsed since the 1972 Moscow summit.

Despite the interruptions in the Agreement due to political considerations, individual project work has continued, often with good success.

Examples of work during the lapsed years include research in 1983 on mutagenesis and environmental chemicals, data gathering on climate change, and an exchange of endangered snow leopards and Przhevalsky horses between zoos in the United States and the USSR.

Evaluating the record

How successful has the Environmental Agreement been? A spectrum of 'success' measures could include at one end complete abatement of environmental damage in each country, and, at the other, satisfaction with an outstanding result from any one project. The answer, therefore, depends on the criteria used to measure success and the evaluation approach, but most reviews of US–USSR exchanges seem to agree that the Environmental Agreement has been one of the most productive in terms of longevity, creation of interaction between scientists, and exchange of information and techniques. For example, a 1985 study by the Kennan Institute, while cautious about some aspects of the agreement, did determine that the Environmental Bilateral offered solid benefits to the United States in terms of communication and knowledge of the Soviet system. Robinson and Waxmonsky, who have strong professional association with the Environmental Agreement, judged the environmental exchange a particularly successful venture because of its yield in scientific research and because of its intent to enhance mutual cooperation. They wrote: 'Today, it is considered to be the most successful of the several cooperation agreements between the U.S. and USSR.' Other scholars of the USSR, including Charles Ziegler and Craig ZumBrunnen have written on the scientific and educational benefits to the United States of the Environmental Agreement.[7]

The cooling-off period in 1980 did interrupt the success of the agreement, as seen in figure 9.1 below, although it seems to be rebounding since 1985. For example, ZumBrunnen cites Waxmonsky as stating that the number of participants in the exchange went from 316 in 1979 to 67 in 1984.[8] The Kennan Report discusses the reduction in projects in 1980 after the Soviet invasion of Afghanistan when exchange visits dropped by 50 per cent.[9]

The Kennan Report also notes that areas which are more policy-oriented were less likely to be active than areas of 'hard science'. For example, the Legal and Administrative Area (XI) and Urban (IV) are described in the report as 'dormant' or 'not yielding tangible results'.[10] One might expect that the record of accomplishments has been uneven, depending on the time period and the area of the agreement, but several areas have had very impressive results. For example, in 1979 under

Table 9.1. *Projects under US–USSR environmental agreements 1972–1987*

Title	Total no. projects	Area share of total (%)	incl. 'Less Active'*	Less active as percent of projects	Area share of active
Air pollution	70	16.7	15	21.4	16.5%
Water pollution	39	9.3	3	7.7	10.8%
Agricultural pollution	28	6.7	6	21.4	6.6%
Urban environment	35	8.3	12	34.3	6.9%
Preservation of nature	121	28.8	39	32.2	24.6%
Marine pollution	19	4.5	1	5.3	5.4%
Biol./genetic conseq.	18	4.3	3	16.7	4.5%
Climate	27	6.4	0	0.0	8.1%
Earthquake	42	10.0	2	4.8	12.0%
Legal & admin.	14	3.3	0	0.0	4.2%
Unclassified	7	1.7	5	71.4	0.6%
Total	420	100.0	86		

* 'Less Active' means no reported activities for project or simple information exchange only.
1985 data unavailable

Area I Project 02.01–12 (Air Pollution: Instrumentation and Measurement Methodology), Soviet and American scientists conducted extensive experiments in Georgia on the formation and transformation of natural aerosols. Modelling techniques, determination of air pollution measurements, and exploration of abatement technologies have been proceeding under the agreement since its inception.

Area IX (Earthquake Prediction) has witnessed a long-term cooperative effort between several US universities, the US Geological Survey, and the all-union and Tadzhik levels of the Academy of Sciences. The effort began in 1973 with an exchange of delegations and a protocol for joint research, and continued with the installation of US equipment in Garm, Tadzhikistan for monitoring seismic action. The Twelfth Memorandum includes further work on this long-term effort, expanding it to include many other projects of earthquake modelling and measurement.

Area V has been the most prolific, accounting for almost 30 per cent of the total projects since 1973 (table 9.1). While some have not yet resulted in substantive output, the variety and vitality of this Area is observable in the appendix listing of projects. Notable accomplishments in Area V include the conclusion of a US–USSR Convention on

Migratory Birds, enhanced breeding of cranes and other endangered species, transfer of animals such as musk-ox, and numerous joint research efforts on marine mammals in the North Pacific.

However, the reader must be cautioned that an objective measure of success may be elusive, particularly in the absence of financial documentation and pollution amelioration evidence.

Five trends notable in the work under the environmental agreement

All of the information discussed below derives solely from the Joint Committee's Implementation Reports rather than from more detailed reports on individual projects. Therefore, the information is limited in its scope and must be understood to offer only a partial picture of the record. In addition, the amount of detail provided in each Implementation Report varies both among years and among Areas.[11]

Despite these limitations, some consistencies of information emerge from the data presented below, which represent nine years of activities and the bulk of projects over the life of the Environmental Agreement thus far.

Information from the reports was coded into ten categories: Meeting Number, Year, Area, Project Number, Project Title (often an abbreviated version), Goals of the project, Activities of the project, Direction of activity (Soviet to US, US to Soviet), Location of Activities, and Responsible Organizations or Agencies. Data was entered for a total of 420 projects, 1,000 activities, and 670 locations. Content of the implementation reports did not allow the systematic collection of information on the number of individual scientists participating in the projects, funding sources, or specific details on publications.

Activities were broken down into eighteen types, from 'no activity' to ones which yielded original publications or actual exchanges of technology.[12] Because of the subjectivity required to code some of the activities from the text of the implementation report, results presented here show endeavours consolidated into seven broad categories: information exchange, visits (counted as a single trip to the other country, not separately for each city visited), research, non-personnel exchanges, original publications, conferences, and development of standards (both scientific and legal).

Locations in the USSR were classified to include Moscow, Leningrad, the Caucasus within the RSFSR, and the fifteen union republics (with the RSFSR other than Moscow and Leningrad divided into European and Asian regions). A category of 'general USSR' was also added when

Table 9.2. *Project Locations in USSR*

Region	Number of projects	Share of total in USSR (%)
Moscow	85	24.9
Leningrad	54	15.8
RSFSR Europe	34	9.9
RSFSR Asia	39	11.4
Ukraine	32	9.4
Kazakhstan	0	0.0
Byelorussia	1	0.3
Moldavia	4	1.2
Estonia	7	2.0
Latvia	3	0.9
Lithuania	1	0.3
Turkmenia	3	0.9
Tadzhikistan	18	5.3
Uzbekistan	4	1.2
Kirgizia	1	0.3
Georgia	14	4.1
Armenia	4	1.2
Azerbaidzhan	4	1.2
Caucasus, Russian	9	2.6
USSR General	25	7.3
Total USSR	342	100.0

locations were not specified or when tours were so widespread that more than seven locations were noted for a single activity.

Participating organizations were broken down into nine types: US federal or state agency, USSR ministry or state committee, university (either American or Soviet), USSR Academy of Science (national level), Union Republic level ministry or committee, American non-governmental research institute, American private business, Soviet enterprise, and miscellaneous.

When the data on project classification were analysed over time, some trends emerged with respect to the questions posed about the work of the Agreement.

Trend 1 Some areas of the agreement have yielded more projects than others consistently over the history of the agreements. The data suggest that most projects have derived from Area V (Preservation of Nature, Nature Preserves), I (Air Pollution), and IX (Earthquake Prediction), while Area II (Water Pollution) has had a large number of projects with at least some activity each year.

As noted in table 9.1, the 420 projects were not evenly distributed among the Agreement Areas. More than half the projects occur in Areas I, V, and IX. Each of the 420 projects represents one project with its own designated code number by the Joint Committee.[13]

When projects are taken out of the list which have been less active in a particular year (no activity registered or only simple information exchange noted), the total number of projects is reduced to 334 and Area V is a bit less dominant, though still with the most projects. Less active projects account for 32 per cent of Area V's projects, but the largest percentage of 'empty nesters' (projects conceived but not yielding any tangible results) is in Area IV (Urban Environment), where 34 per cent of the thirty-five projects initiated over the period studied did not produce activity.

This result suggests that not all areas of the environmental agreement are equal; some have been more successful in terms of action, while others have been quite dormant.

Trend 2   The Slavic Core areas (Russian and Ukrainian republics) dominate the geographic distribution of activities under the Agreements. Project locations are shown in table 9.2, noted only for projects which had activity other than information exchange and for which a location was described in the implementation reports (total of 243 projects). Just as Washington DC dominated US activities, the 'headquarters' location of Moscow dominates the Soviet side and, in fact, has twice the number of activities as Washington DC, Leningrad and Moscow together which account for 41 per cent of the total Soviet locations described in the accords. While the Russian republic as may be expected is the main location with 64 per cent of the USSR total, the Ukraine, Tadzhikistan and Georgia have also been the scene of numerous projects, and most union republics have had at least one activity. A surprising omission is Kazakhstan, which has not shown up in the nine agreements examined as the location for any project activity, although it may have hosted some projects without being noted in the reports.

This concentration raises issues about the possible dispersal of activity, should independent states arise out of the Soviet Union. Would the United States seek to establish separate environmental accords with newly independent republics? The geographic concentration of activities may reflect the dominance of the core in terms of scientific personnel and facilities, ease of contacts with Westerners, and even allocation of research funding.

Trend 3   Active projects have tended to go through a cycle of maturation, with information and personnel exchanges in the early phases, followed by research or exchange activities, and production of

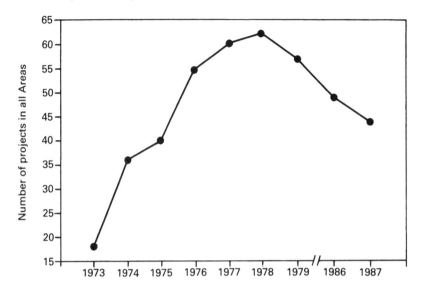

Figure 9.1   US–USSR Agreement Projects

publications. 1979 was a peak year for activities, followed by a decline during the lapse phase of the Agreement. The current period seems to represent a new period of building.

The data on activities and number of projects shown in figures 9.1 and 9.2 suggest that 1978–9 did represent a peak period when many projects were either completing work initiated in the early 1970s or continuing to make progress. While the Eighth report is not included here, the 1986 Ninth Meeting report did show the drop-off of activities after 1980 which is noted in the Kennan document.

Information exchanges and visits by scientists tend to be the most common activities, but their share of all activities has shrunk over time as the Environmental Agreement matures and work proceeds into other areas such as joint research. For instance, in 1973 86 per cent of the activities for the first year were in these two categories, but by 1979 the share had declined to 49 per cent (figure 9.3). Likewise, original publications from the projects increased over the same period, although a decline occurred for 1987 even as the share of information-visit activities increased again that year. These facts may represent a 'retrenchment' in 1987 with the initiation of new activities and projects, culminating in the many new added projects for the 1990 memorandum.

Research and non-personnel exchange activities have been more

134 Kathleen E. Braden

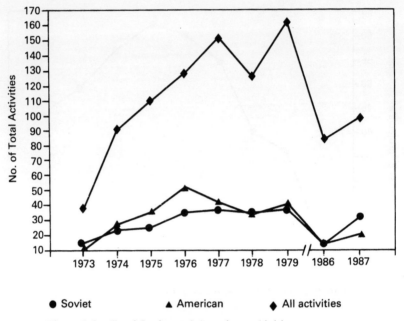

Figure 9.2 Total Soviet and American activities

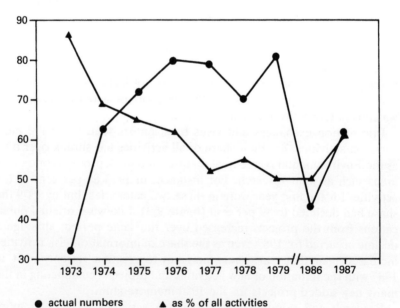

Figure 9.3 Information exchange and visits

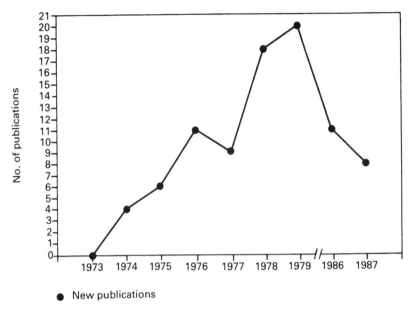

● New publications

Figure 9.4 Publications from projects

erratic over the period studied (figure 9.4) although for research, an upward trend existed until 1979, a year when many project activities and reports were completed.

How reciprocal are the exchanges? *Trend 4* shows that except for the area of hardware technology exchange, the Environmental Agreement has been quite reciprocal in terms of *quantities* of activities.

Figures 9.5 and 9.6 compare Soviet and American participation in the Environmental Agreement. Activities were coded as originating with the Soviet or American side when specified in the reports. A remarkable fact is that over the period examined, only an 8 per cent difference exists in the total number of activities by each side. Visits are particularly reciprocal, with Soviet visits (representing a project delegation, not individuals) totalling 215, and American 216! Also striking, however, is the imbalance in hardware technology exchange, a rather limited portion of the Agreement. American export of equipment totalled 90.4 per cent of the twenty-one total activities in this category, largely earthquake monitoring and air pollution monitoring equipment. On the other hand, Soviet export of samples to the United States outnumbered US samples to the USSR by almost two-to-one; this represents either samples brought back from the USSR by US scientists or samples sent for analysis.[14]

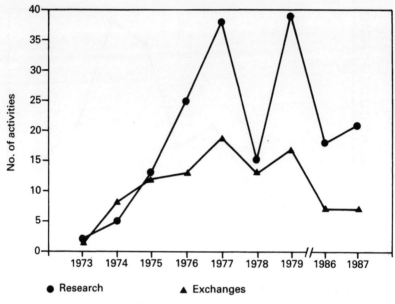

Figure 9.5 Research and non-personnel exchanges

Figure 9.6 Soviet and American visits

US–Soviet cooperation for environmental protection 137

Field research, general collaborative research, and joint experiments appeared to be fairly reciprocal. Thirty activities in this category were reported for Soviet scientists in the US, and forty for US scientists in the USSR.

Finally, the type of participating organisations for the various projects was described. *Trend 5* demonstrates that the Environmental Agreement has been government- and academia-dominated.

The major government agencies such as EPA and Hydromet and the Academy of Sciences tend to be the most active players. Universities also share in the work, but private US business, non-governmental institutes and Soviet enterprises are rarely noted. The all-union level tends to be the most common Soviet factor, although some republic level agencies, such as in Tadzhikistan and Estonia, have occasionally been represented. Should diversification, particularly on the Soviet side, become more desirable, this concentration of activities into certain sectors will have to be addressed.

Conclusions

The formation of Goskompriroda, the State Committee on Nature, was accomplished by the USSR in 1988, and it has control over international environmental agreements. New legal frameworks are being created to govern amelioration of adverse environmental impacts in the USSR.[15] On the other hand, this new 'super-agency' has ironically been established in a period when a decrease in central authority over the environment is occurring in two ways: (1) more control, especially responsibility for financing amelioration and preservation, is being allocated to the local level as union republic governments demand more autonomy; and (2) more citizen-based, non-governmental environmental organisations are emerging.[16]

Thus, the Environmental Agreement as it enters its period of rebuilding is faced with change and opportunity. The Twelfth Memorandum for 1990 already suggests that the breadth of agencies responsible for the projects is increasing on both sides, and more issues of a global nature are tackled (see appendix). Funding issues are still plaguing the Joint Committee, as noted by Robinson and Waxmonsky, but one can hope that increased environmental concern on the part of the public in both nations may help ease the shortfall.

An interesting issue facing the Joint Committee will continue to be the interplay of economics, policy, and environmental quality. The legal questions associated with making the polluter pay are emerging in the USSR, and in a time of economic emergency the environmentally-unwise but economically-attractive option may yet win out. F. Morgun warns against allowing the need for foreign hard currency to bring in 'dirty' companies which exploit the Soviet environment and damage its ecology.[17]

The Environmental Agreement has worked most productively when the projects remain in the realm of natural scientists. Area IV and Area XI have been less successful in dealing with social institutions, yet these appear to be the very themes that will require much attention in the future. An encouraging sign is the proliferation of activities in the Twelfth Memorandum under Areas IX and VII (Biological and Genetic Effects of Environmental Pollution). For example, while Philip Pryde and others have recommended that the USSR needs to adopt a type of environmental impact evaluation procedure, the Twelfth meeting included a project under Area IX (02.11–1103) on Environmental Impact Assessment.[18]

Finally, a difficult component to measure, but one that may be crucial to the success of the Environmental Agreement, is the human factor. This author noted that many of the same names of scientists appeared year after year in the reports, indicating an interest in seeing projects completed. The long-term pairing of US and Soviet scientists and policy makers, overcoming linguistic, financial, and cultural barriers may be an important determinant of success. Similarly, the government employees who manage the Joint Committee, often on little funds and staffing, have been able to accumulate a respectable record of cooperation and results. Many of their remarkable accomplishments have been on national-level issues, under projects coordinated by large, government agencies in the area of hard science. It will be a challenge to move their endeavours to the global scale, to the decentralised administrative level, and to the realm of policy, but it is a process they have already begun.

If the many years of cooperative work on the environment have been of benefit in transferring Western know-how to the USSR, the key may therefore be not in knowledge embodied in technique and machinery, but in ways of thinking and viewing solutions to environmental problems. This is a type of knowledge embodied in the human mind, in fruitful contacts with colleagues, and in exposure to cultural factors involved in science and policy-making. It may prove to be the most successful convergence point of two disparate systems.

Notes

1. See, for example, the classic work by Marshall Goldman, *The Spoils of Progress: Environmental Pollution in the Soviet Union* (Cambridge, MA: MIT Press, 1972). For an example of recent releases of new data on environmental disruption, see: Goskomstat SSSR, *Okhrana okruzhayushchei sredy i ratsional'noye ispol'zovaniye prirodnykh resursov v SSSR* (Protection of the Environment and Rational Use of Natural Resources in the USSR) (Moscow: Finansy i statistika, 1989).
2. Eisenhower World Affairs Institute, 'US–Soviet Exchange: The Next Thirty Years', Conference Report, Gettysburg, Pennsylvania, 28 January–1 February 1988, pp. 37–42.
3. Kennan Institute for Advanced Russian Studies, 'US–Soviet Exchanges', Conference Report, Washington, DC, 1985, Part II.
4. United States Congress, Committee on Science and Technology, US House of Representatives, Subcommittee on Domestic and International Scientific Planning and Analysis, *Background Materials on US–USSR Cooperative Agreements in Science and Technology*, Washington, DC, GPO, November, 1975, 'Cooperation in Environmental Protection', p. 23.
5. Nicholas Robinson and Gary Waxmonsky, 'The US–USSR agreement to protect the environment: 15 years of cooperation', *Environmental Law*, 18 (1988), pp. 403–447; 403.
6. US–USSR Committee on Cooperation in the Field of Environmental Protection, 'Memorandum of the Twelfth Meeting of the US–USSR Joint Committee on Cooperation in the Field of Environmental Protection', Washington, DC, 9–12 January 1990, p. 88.
7. Kennan Institute, *US–Soviet Exchanges*, p. 29; Robinson and Waxmonsky, 'US–USSR agreement', p. 406; Charles Ziegler, *Environmental Policy in the USSR* (Amherst, Mass: University of Massachusetts Press, 1987); Craig ZumBrunnen, 'Gorbachev, economics, and the environment', in US Congress, Joint Economic Committee, *Gorbachev's Economic Plans*, vol. 2, November 1987, Washington, DC, GPO, pp. 397–424.
8. ZumBrunnen, ibid., 'Gorbachev', p. 417.
9. Kennan Institute, *US–Soviet Exchanges*, p. 28.
10. Kennan Institute, *US–Soviet Exchanges*, pp. 23–4.
11. The author was handicapped by the absence of data from the 1985 and 1988 meetings of the Joint Committee (documents unavailable). Thus, project accomplishments of the Eighth and Eleventh Meetings are not included here. Additionally, several individual sections on specific projects (such as commercial fur-bearer activities and one year of the crane study) were omitted from the documents available to the author, and the 1987 Implementation Report was obtained in draft form only.
12. The author classified activities as follows: No Activity; Simple Information Exchange; Exchange of Personnel (subdivided into visit to institution, visit to research site or field facility, conference attendance); Collaborative research (type not specified); Modelling; Original Publication; Field

Research; Joint Experiments; Exchange of Samples; Exchange of Organisms; Exchange of Hardware-Technology; Exchange of Systematic Publications (journals, papers); Development of Educational Materials; Development of Scientific Standards; Development of Administrative Standards or Domestic Law; Development of Measurement Techniques; Development of International Agreement; Support for Citizens' Organizations.
13. Many are sub-projects under larger endeavours. In cases of sub-projects, the main project is not registered so as to avoid duplication of activity count. Also, each of the 420 represents one project in one year; thus, 'Stationary Source Air Pollution Control Technology' is counted many times because it appears as a separate project in various years' implementation reports.
14. For instance, in 1979 under Area V Project 02.05-2101 US scientists collected several hundred examples of soil invertebrates in Magadan Oblast.
15. See for example, Gosudarstvennyi komitet SSSR po okhrane prirody, *Doklad: Sostoyanie prirodnoi sredy v SSSR v 1988 godu* (Paper on the USSR Environment in 1988) (Moscow, 1989). F. Morgun, 'Ekologiya v sisteme planirovaniya (Ecology in a System of Planning), *Planovoye khozyaistvo*, no. 2 (1989), pp. 53–61, as excerpted in: *Environmental Policy Review*, 3, no. 2 (July, 1989), pp. 16–22. Andrei Yablokov, 'The current state of the Soviet environment', *Environmental Policy Review*, 4, no. 1 (January 1990), pp. 1–14.
16. United States Congress, Commission on Security and Cooperation in Europe, *Politics of Pollution in the Soviet Union and Eastern Europe*, 26 April 1988, Washington, DC, GPO. (Testimony of Barbara Jancar).
17. Morgun, 'Ekologiya', pp. 11–12.
18. Philip Pryde, 'The future environmental agenda of the USSR', *Soviet Geography*, XXIX, no. 6 (June 1988), p. 560. US–USSR Committee on Cooperation in the Field of Environmental Protection, *Memorandum of the Twelfth Meeting of the US–USSR Joint Committee on Cooperation in the Field of Environmental Protection*, Washington, DC, 9–12 January 1990, p. 86.

Appendix: Projects from 1973, 1978, 1986, and Planned projects, 1990

Notes: Some project titles here are shortened for conciseness.
'New projects 1986' were added at end of report and often not assigned to special Area at time of initiation; follow-up 1987 implementation report indicated that new projects of 1986 were beginning to yield resulting activities.
Projects under Area X, Arctic and SubArctic Ecological Systems were traditionally allocated to other areas in 1973, 1978, and 1986.

I The Prevention of Air Pollution

1973

Air Pollution Modelling, Instrumentation, and Measurement Methodology
Stationary Source Air Pollution Control Technology
Transportation Source Air Pollution Control Technology

1978

Air Pollution Modelling and Standard Setting
Instrumentation and Measurement Methodology
Development of Wet Limestone Method
Gas Desulphurisation by Magnesia Method
Gas Desulphurisation by Ammonia Scrubbing
Particulate Abatement Technology
Demetallisation Pretreatment During Hydrosulphurisation of Heavy Residual Oils to Produce Low Sulphur and Low Ash Fuel Oils
Complex Methods of Fuel Usage (Coal and Oil) in Energy Generating Systems with Prevention of Harmful Emissions into the Environment
Coal Flotation Desulphurisation Technology
Protection of the Environment from Influence of Coal Processing Plant Operations
Ferrous Metallurgy Pollution Control Technology
Transportation Source Air Pollution Control Technology

1986

Air Pollution Modelling and Standard Setting
Instrumentation and Measurement Methodology
Wet Limestone Technology
Gas Desulphurisation by the Magnesia Method
Gas Purification by Ammonia Scrubbing/Alkali Scrubbing
Particulate Control
Methodology for Determining the Physical and Chemical Properties of Aerosols
Protecting the Environment from Effects of Coal Cleaning Waste
Ferrous Metallurgy Pollution Control Technology
Transportation Source Air Pollution Control Techniques

1990

Air Pollution Modelling and Standard Setting
Instrumentation and Measurement Methodology
Remote Sensing of Atmospheric Parameters
Stationary Source Air Pollution Control Technology
Removal and Suppression of Oxides of Nitrogen from Stationary Sources
Gaseous Organic Emissions Abatement from Stationary Sources
Wet Desulphurisation Processes
Dry Desulphurisation Processes
Municipal Waste Processing and Disposal

II The Prevention of Water Pollution

1973

Studies and Modelling of River Basin Pollution
Protection and Management of Lakes and Estuaries
Pollution and Aquatic Ecological Systems
Prevention and Treatment of Waste Discharge

1978

River Basin Water Quality Planning and Management
Protection and Management of Water Quality in Lakes and Estuaries
Effects of Pollutants Upon Aquatic Organisms and Ecosystems
Development of Water Quality Criteria
Prevention of Water Pollution from Municipal and Industrial Sources

1986

River Basin Water Quality Planning and Management
Water Quality in Lakes and Estuaries
Effects of Pollutants on Aquatic Organisms and Ecosystems
Prevention of Water Pollution from Municipal and Industrial Sources

1990

River Basin Water Quality Planning and Management
Water Quality in Lakes and Estuaries
Effects of Pollutants on Aquatic Organisms and Ecosystems
Effects of Pollutants on Wetlands Ecosystems
Anthropogenic Impacts on Ecosystems of Reservoirs and Lakes

Prevention of Water Pollution from Municipal and Industrial Sources
Control of Water Pollution from Pulp and Paper Enterprises
Groundwater Quality

III Environmental Pollution Associated With Agricultural Production

1973

Pollution and Agricultural Production

1978

Integrated Pest Management
Interaction Between Forests, Plants, and Pollutants
Forms and Mechanisms by Which Pesticides and Chemicals Are Transported
Effects of Chemicals Used in Agriculture on Fauna

1986

Interaction Between Forest Ecosystems and Pollutants
Forms and Mechanisms by Which Pesticides and Chemicals Are Transported

1990

Air Pollution Effects on Vegetation, including Forest Ecosystems
Forms and Mechanisms by Which Pesticides and Chemicals are Transported

IV Enhancement of the Urban Environment

1973

Enhancement of Urban Environment

1978

Urban Transportation and the Environment
Historic Monuments
Solid Waste in Urban Areas

Enhancement of Environment in Existing Cities Through Urban Land Use
Recreation Zones in Urban and Near Urban Areas

1986

Urban Transportation and The Environment
Heritage Resources Conservation and Management
Solid Waste in Urban Areas

1990

Heritage Resources Conservation and Management
Research Conservation and Management of the Beringian Heritage
Natural, Historical, Urban Development and Historical Archival Heritage
Solid Waste in Urban Areas

V Preservation of Nature and Organization of Preserves

1973

Preservation of Nature and Organization of Preserves

1978

US–USSR Convention on Migratory Birds
Black-Footed Ferret
Raptors
Cranes

Study of Captive Propagation of Fur-Bearers
Steppes, Prairies, Floodplains, Deltas
Protected Natural Areas
Publication of Joint US–USSR Compilation of Articles on Principles of Conservation of Wildlife
Development of Exchange of Scientific and Popular Information
Immobilization
Rational Utilization and Production of Wildlife
Censusing Wildlife Population and Estimating Sustainable Yield
Ungulates in Arctic and SubArctic
Taiga-Tundra

Permafrost-Related Problems
Reclamation and Revegetation of Disturbed Land
Biosphere Reserves
Arid Ecosystems
Marine Mammals
Threatened Plants and Introduction of Exotic Species
Northern Migratory Waterfowl
Ecology and Economic Significance of the Wolf
Mammals of Holarctic

1986

Implementation of US–USSR Convention on Conservation of Migratory Birds and their Environment
Cranes and Other Rare Birds
Rare and Endangered Animals
Protection of Natural Areas and Ecosystems
Cooperation Among Zoos in Captive Breeding of Rare and Endangered Animals
Biosphere Reserves
Arid Ecosystems
Marine Mammals
Threatened Plants and Introduction of Exotic Species
Northern Migratory Waterfowl
Holarctic Mammals
Ichthyology and Aquaculture

1990

Implementation of US–USSR Convention on Conservation of Migratory Birds and their Environment
Cranes and Other Rare Birds
Rare and Endangered Animals
Protection of Natural Areas and Ecosystems
Cooperation Among Zoos in Captive Breeding of Rare and Endangered Animals
Biosphere Reserves
Arid Ecosystems
Marine Mammals
Threatened Plants and Introduction of Exotic Species
Northern Migratory Waterfowl
Holarctic Mammals

Chemical Senses and Communication in Animals
Contemporary Technology and Ecological Studies of Large Mammals
Ichthyology and Aquaculture

VI Marine Pollution

1973

Marine Pollution – Shipping
Pollution and Well Drilling – Pipeline Transportation
Pollutants and Marine Organisms

1978

Prevention and Clean Up of Oil Pollution from Shipping
Effects of Pollutants on Marine Organisms

1986

Prevention and Clean Up of Pollution of Marine Environment from Shipping
Effects of Pollutants on Marine Organisms

1990

Prevention and Clean Up of Pollution of Marine Environment from Shipping
Transport, Partitioning, and Effects of Radioactivity Releases in a Marine Ecosystem

VII The Biological And Genetic Effects of Environmental Pollution

1973

Pollution and Biological/Genetic Consequences – Analysis of Environment
Biological and Genetic Effects of Pollutants

1978

Biological and Genetic Effects of Pollutants
Comprehensive Analysis of the Environment

1986

Biological and Genetic Effects of Pollutants Mutagenesis
Comprehensive Analysis of the Environment

1990

Biological and Genetic Effects of Pollutants Mutagenesis
Comprehensive Analysis of Environmental Protection Measures
Climate Change Impacts and Responses
Marine Ecosystems and Ecological Problems of World Ocean
Prediction of Ecological Impacts of Environmental Pollutants

VIII   The Influence of Environmental Changes on Climate

1973

The Influence of Environmental Changes on Climate

1978

Effects of Changes in Heat Balance of the Atmosphere on Climate
Effects of Pollutants of the Atmosphere on Climate
Changes in Solar Activity and Climate

1986

Effects of Changes in Heat Balance of Atmosphere on Climate
Effects of Pollution of Atmosphere on Climate
Influence of Changes in Solar Activity on Climate

1990

Climate Change
Atmospheric Composition
Radiative Fluxes, Cloud Climatology, and Climate Modelling
Data Exchange Management
Stratospheric Ozone

IX Earthquake Prediction

1973

Earthquake Prediction

1978

Field Investigation of Earthquake Prediction
Physics of the Earthquake Source
Mathematical and Computational Prediction of Places Where Earthquakes Occur
Engineering-Seismological Investigation
Simultaneous Tsunami Warning System

1986

Earthquake Prediction
Engineering–Seismological Investigations
Field Investigation of Earthquake Prediction
Laboratory and Theory Investigation of Physics of Earthquake Source
Theoretical Models and Algorithms for Earthquake Prediction
Simultaneous Warnings on Tsunamis

1990

Earthquake Prediction
Laboratory and Theoretical Investigations of the Earthquake Source
Laboratory Studies of Rock Mechanics
Numerical Modelling of Active Earthquake Faulting
Phenomenogical and Theoretical Models for Earthquake Prediction
Theoretical Models in Specific Systems of Seismo-Active Faults
Worldwide Tests of Existing Algorithms
Seismic Emission
Slow Wave Studies
High Frequency Studies
Vibro Sounding Studies
Simultaneous Warnings on Tsunamis
Seismological Studies and Data Exchange

X Arctic and Subarctic Ecosystems

1990

Oil and Gas Developments in Permafrost Regions
Regional Developments of Northern Landscapes and Prevention of Negative Impacts
Ecological and Cultural Studies in Northern Landscapes
Monitoring of Cryospheric Processes
Preservation of Arctic Ecosystem in Bering Region

XI Legal and Administrative Measures for Protecting Environmental Quality

1973

Legal and Administrative Measures for Protecting Environmental Quality

1978

Legal and Administrative Measures
Harmonization of Air and Water Pollution Standards

1986

Legal and Administrative Measures

1990

Policy, Law, and Management in the Area of Environmental Protection
International Environmental Law and Policy
Environmental Impact Assessment

New Projects 1986

Education and Training in Field of Environmental Protection
Groundwater Quality
Developing and Perfecting Waste Free and Low Waste Technologies for Preventing Environmental Pollution
Processing and Utilizing Harmful Toxic Wastes

New Project, 1990

Education and Training in the Field of Environmental Protection

10 US–Soviet nuclear safety cooperation: prospects for health and environmental collaboration

Michael B. Congdon

Despite the initial difficulty in getting accurate information on the Chernobyl accident and its transboundary radiation effects, US policy during the following months sought to avoid placing the Soviets on the defensive. Rather, the United States almost alone among Western nations hoped to encourage openness on the part of the Soviets to Western ideas about nuclear safety. This approach has ultimately paid rich dividends in encouraging technical exchanges of mutual benefit to both sides and to the rest of the world. Indeed, if Chernobyl has a silver lining, it is the effect it has had on changing the way the Soviets have dealt internationally with a domestic disaster. Secrecy has given way to cooperation in one small, but vital, technological area of US–Soviet relations.

On 26 April 1988, the US Nuclear Regulatory Commission signed an agreement on a programme of cooperation with the Soviets in Civilian Nuclear Reactor Safety. This set in motion a programme of extensive cooperation including a number of firsts in US–Soviet technical cooperation. The US agreed to a ten-point programme which is one of the most extensive technology transfer agreements in US–Soviet history and one of the most extensive looks the US has into state-of-the-art Soviet technology.[1] This agreement was the culmination of over two years of effort seeking to open up the Soviet Union to such cooperation.

US policies have led to extensive exchanges in nuclear safety technology and regulation, but results in the areas of environmental and health effects of the Chernobyl accident have not yet been as successful. Soviet scientists stress how the magnitude of the tasks they face treating the many victims of this accident prohibits pure scientific exchange which may have long-term benefits but does little to help the immediately affected. They favour cooperation with the US, but indicate they have no time for theoretical studies. US visitors report that organisations compete for Western resources, bureaucratic problems abound, and they are no longer dealing with a centralised Soviet structure.

The prospect of further collaboration is threatened as the issue of the health and environmental effects of Chernobyl has become politicised. Environmental protests in the USSR continue to proliferate, based on new revelations about inadequate attention to the health needs of the population which continues to live in the contaminated areas. Soviet responses to US proposals on health effects were, until recently, focused on obtaining US financial assistance. And cooperation with the US in general on nuclear safety can be seen as an attempt to gain a Western 'seal of approval' on their handling of the accident's consequences. Instead of engaging the US in cooperation, for example, the Soviets turned to the International Atomic Energy Agency for an 'independent analysis' of their strategy for protecting the population from the effects of radiation – but on Soviet terms, and using Soviet data. Only in preparations leading to the June 1990 Bush–Gorbachev Summit have the Soviets finally shown the flexibility to cooperate bilaterally with the US in the health-effects studies which the West considers so vital.

Theoretical background

Issues such as protection of the environment, limiting illegal drug sales, non-proliferation, and economic development raise questions concerning how scientific and technological developments should influence a broad range of foreign policy decisions. They emphasise how few contemporary foreign policy issues are addressable from the standpoint of traditional state relations. Moreover, the linkages among such issues are highly complex, often involving competing interdependencies. At the centre of many are critical questions of technological progress as applied to political values, frequently decreasing the ability of the nation state to control its own destiny.

Nuclear safety is one in a category of 'global issues' that cross political boundaries of nation states. These issues are bound to increase as the Cold War between the US and the USSR recedes into history. The literature of modernisation lays a disciplinary groundwork for studying why 'global issues' have come about.[2] Nations increasingly deal with complex issues that blur distinctions between the policy sciences and the technical sciences. Their interdependence leads policy makers to stop seeking purely national solutions and instead to seek cooperative, multilateral solutions.

Rather, such issues require international approaches which have been termed 'régimes', which often operate across unfamiliar boundaries.

These may involve formal agreements, but they are more likely to involve shared norms, rules, and patterns of behaviour, which can be made applicable to all parties.[3] Régimes may imply, for example, agreed limits on certain types of resource exploitation, such as in the well-known theory of the commons, as well as patterns of behaviour that require adherence to terms that go beyond the interests of any single state, so long as others are similarly willing to limit their actions.

The Chernobyl experience is unique in modern history. It came at a time of fresh openings in the relationship between the two great power blocs, and these openings were effectively exploited by the West. Chernobyl was treated in US policy as a 'global issue', rather than a Soviet failure, and this perspective has resulted in a dramatic increase in US–Soviet cooperation. This strategy was consistent with the approach to international relations that is based on interdependence and mutually beneficial solutions to technology-based international problems. As a result of this policy, what had been a human tragedy of almost mythic proportions has led to extensive technological cooperation.

But Soviet domestic politics, brought about as a result of glasnost and perestroika has limited the Kremlin's willingness to extend this cooperation into the health-effects area, although progress is being made both bilaterally and through the International Atomic Energy Agency. It is the thesis of this paper that collaboration needs to expand into the area of the health and environmental effects of the Chernobyl radiation. It is not yet clear whether the Soviets are prepared to engage in serious technical cooperation by sharing critical information in this politically explosive area. Such cooperation, however, would be recognised as true glasnost by the international science community, and could serve as a model for further technical exchanges with the nations of Eastern Europe, leading to East–West interdependence which can only benefit their economies and societies.

The evolution of US–Soviet bilateral nuclear safety cooperation

As early as the summer of 1984, before liberalisation had begun in the USSR, US President Ronald Reagan proposed a new policy on scientific and cultural exchanges with the Soviets, encouraging Federal agencies to consider expanding their level of activities that had been curtailed as a result of the Soviet invasion of Afghanistan in 1979. The policy was to support improved communications with the Soviets, and it was a major shift in the Reagan Administration's approach to US–Soviet relations.

As one element of this policy, the US Nuclear Regulatory Commission (NRC) initiated contact with the Soviets in late 1985 in order to open discussions on nuclear safety matters. By early 1986, the Soviets had approved 'in principle' a US NRC delegation to the USSR, but little progress was made on scheduling a visit. The Department of State at the time believed that the Soviets were holding the US at arm's length on any substantive exchanges in this area of advanced technology which the Soviets had always treated as a state secret.

Separate from the NRC initiative, the Soviet Embassy in Washington formally proposed, in February 1986, the convening of a sixth Joint Committee Meeting (JCM) in the USSR under the US–Soviet Agreement on the Peaceful Uses of Atomic Energy.[4] Shortly before Chernobyl, the US Department of Energy (DOE), which administers that Agreement for the US, accepted the Soviet proposal and suggested an agenda covering fusion, breeder reactors, and high energy physics,[5] but not civilian nuclear reactor safety. The Chernobyl accident on 26 April halted plans for an NRC trip in the period immediately following, but it did not affect planning for the JCM. As a result of the accident though, DOE proposed adding light-water reactor safety to the discussions at the JCM. The Soviets accepted this proposal, and the parties agreed to meet in Moscow in August 1986.

Also in August, the International Atomic Energy Agency (IAEA) held a major international conference on the Chernobyl accident. This 'Post-Chernobyl Review' turned out to be one of the most significant events in turning the Soviets towards international cooperation on nuclear safety and away from a purely national approach to this problem. The United States sent a delegation of twenty-one experts, led by senior officials of the US Departments of Energy and State and the NRC. A number of private and academic experts also participated for the US. Early in the week during the conference, US experts, frustrated by the lack of time allowed for detailed questioning of the Soviet participants in the plenary sessions, sought agreement from the delegation leadership to meet privately with Soviet experts. US leaders set up the meetings in conversations with the Soviet political leadership. US and Soviet technical experts thereupon met in private early morning sessions on the three days prior to the opening of the daily plenary meetings.

US delegation members reported surprisingly frank and open exchanges of information with their Soviet counterparts and obtained their commitment to answer a list of as many questions as the US side wished to propose following the meeting. It became clear that the large Soviet delegation, many of whom were scientists who had not previously been allowed to travel to international conferences, were operating under instructions to be open and candid about the causes and effects of the accident with their scientific peers, including the American delegation.

Upon their return from Vienna, US delegation leaders recommended expanded cooperation with the Soviets both in bilateral exchanges and in the IAEA. They recommended joint work leading to enhanced safety in the following areas:

> improving operational safety through operator training, emergency procedures, use of accident simulators, and safety accreditation;
>
> decontamination, decommissioning, clean-up, entombment, and any other actions that proved necessary to rehabilitate the area around the reactor that was affected by the accident; and
>
> detailed epidemiological studies to improve common knowledge of low-level radiation health effects.

At the same time, the US urged the Soviets to undertake additional safety studies using computer-based probabilistic risk assessment techniques (pioneered in the US after the Three Mile Island accident in 1979) in order to assure a comprehensive identification of their reactors' safety vulnerabilities.

While the Soviets had refused to accept US offers of assistance in the weeks immediately following Chernobyl, the stress of coping with the aftermath of the accident and the extreme interest expressed by participants at the Vienna conference appeared to heighten their sensitivity to cooperation in nuclear reactor safety. The US indicated that it wished to share its experience in implementing safety improvements during the years following the 1979 Three Mile Island nuclear reactor accident, the effect of that experience on the development of current safety technology, and US safety philosophy and standards. The NRC representative at the August 1986 JCM outlined NRC's role and responsibilities, and its programmes of international cooperation in nuclear safety. He also restated NRC's interest in leading a delegation to the USSR to visit Soviet facilities and to discuss safety matters, and the Soviets agreed to extend an invitation.

Accordingly, in March 1987, the first US nuclear safety team, led by an NRC Commissioner, visited the USSR and toured a number of Soviet facilities, including Chernobyl Units 1 and 2, the Izhora reactor components production plant near Leningrad, the Zaporozhe atomic power station, and the fast breeder reactor near Beloyarsk. They concluded their trip in Moscow with discussions on the safety and economics of nuclear plant operations. Each site visit was valuable in revealing how inadequate safety standards and practices were in the

USSR at the time of Chernobyl. At each stop the delegation discussed four broad technical areas of concern to the US, namely: nuclear safety regulation, policy, and practices; safe operations; safety research; and radiation protection and health effects.[6]

The NRC (1) emphasised that the Soviet approach to safety in their latest pressurised water reactors (PWRs) was comparable to those in other countries, but (2) criticised the implementation of this approach in operating their older reactors. The team also stressed that the medical treatment of the victims of the accident was of great interest to the US. In this regard, seven of the US officials, who travelled to Kiev but were not invited to Chernobyl, held instead a two-hour discussion with eleven Soviet radiation health specialists including the Deputy Minister of Health Care at the Ukrainian Ministry of Health headquarters.

The Soviets at the Kiev meeting responded openly to US questions, but did not seem eager to enter into extensive discussions about cooperation. They described the very difficult situation after Chernobyl, with health care personnel working day and night to cope with the severe health problems of the victims. They described the immediate actions taken to prevent infectious diseases and reported that extensive food and water monitoring was implemented and that all food products outside the 30-kilometre exclusion zone around the plant were suitable for consumption except for some dairy products and certain berries. Special attention, they said, had been paid to pregnant women and to children; all of the former were evacuated the first day.[7]

A Protocol was signed at the end of the visit that stressed the possibility of mutual interests in the four broad technical areas mentioned above and promised a return visit by the Soviets to the United States. The reciprocal visit would, in the words of the Protocol 'provide an opportunity for further exchanges of views and safety experiences by the two sides, and allow the Soviet delegation to visit several corresponding nuclear facilities and institutions in the US'.

In late September 1987, the IAEA hosted another meeting in Vienna on severe accidents. This meeting provided important evidence that the Soviet Union and Eastern Europe were beginning to open doors for their scientists to participate in international conferences, particularly on nuclear safety matters. Nearly forty Soviet scientists and nuclear officials attended, most of whom delivered papers.[8] Less than two weeks after the US delegation returned from this meeting, fourteen Soviet safety experts and officials, some of whom had also participated at Vienna, arrived in Washington for a two-week stay in the US to visit US nuclear facilities. The Soviet delegation included the Deputy Minister of Atomic Power, representatives of two major state committees and the

Kurchatov Institute, and veterans of the Chernobyl clean-up. The US host was NRC Chairman Lando Zech.[9]

The head of the Soviet delegation, Deputy Minister of Atomic Power Alexander Lapshin, offered a far-reaching proposal for a ten-year agreement on cooperation in nuclear safety regulation, research and methodology, and reactor design, construction, and operation. Significantly, health and environmental effects were *not* mentioned. The US responded with a counter-draft in January 1988. It designated the NRC as lead agency for the US on nuclear safety cooperation under the US–USSR Agreement on the Peaceful Uses of Atomic Energy, and it proposed setting up a separate Joint Coordinating Committee on Civilian Nuclear Reactor Safety (JCCCNRS) under this Agreement.

US objectives entering the negotiations were, firstly, to obtain Soviet agreement to cooperate on nuclear reactor safety as a means to enhance the safety of Soviet reactors. Secondly, the US wanted to explain Western nuclear safety philosophy to help induce the Soviets to adopt a 'defence in depth' approach to the design and operational safety of their nuclear power plants. Thirdly, the US wanted cooperation that could help obtain critical information about the Chernobyl accident itself and its environmental consequences and health effects. Fourthly, the US hoped to involve the US nuclear industry in the hoped-for market for nuclear technologies and safety-related services in the Soviet Union and Eastern Europe.[10]

The Memorandum of Cooperation was signed in Washington on 26 April 1988, exactly two years after the accident. It established a joint coordinating committee (the JCCCNRS) to facilitate technical exchanges in the same four basic principles that motivated the NRC's original trip. The Soviets agreed that the JCCCNRS would first meet in the USSR to begin cooperation, and Chairman Zech accepted a Soviet invitation to lead the US delegation. Zech asked his top civil servant, the Executive Director for Operations (EDO), to make resources available for expanded cooperation with the Soviets, and the Deputy EDO was appointed the US co-chairman.

In all, seventeen persons made the trip in August 1988, including State and Energy Department officials. They were the first non-Soviets to visit the entombed Chernobyl Unit 4 reactor and sarcophagus. The ensuing Protocol of this visit said that both sides 'look forward to continued enhancements of safety through greater understanding of each other's approaches and to making further improvements in the safety of their respective plants'.[11] Chairman Zech also met with the leaders and other senior officials from the Soviet organisations working on ensuring nuclear power safety, in particular Chairman Vadim Malyshev of the

State Committee for Supervision of Nuclear Power Safety and Minister Nikolai Lukonin of the Ministry of Atomic Power. Also in August 1988 the Protocol of the First Meeting of the JCCCNRS laid out plans which had been negotiated over the summer for cooperation in ten technical areas resulting in ten Working Groups.[12]

Project milestones: The visit of Vadim Malyshev and the inspector exchange

Two milestones in developing confidence between the two sides occurred in 1989 when Vadim Malyshev, then Chairman of the Soviet Nuclear Regulatory Commission (GAEN), visited the United States at the invitation of NRC Chairman Zech[13] and two US NRC resident nuclear power plant inspectors traded places with two Soviet inspectors for seven-week tours at reactors in the other country. Advance planning for the Malyshev visit called for discussions of decommissioning,[14] safety aspects of first generation plants, seismology (the recent earthquake in Armenia was on everyone's mind), diagnostics and controls (advanced technology or computerised control rooms), the handling of public opinion, perfecting regulatory techniques and practices, and details for the proposed inspector exchange. NRC took Malyshev to the Calvert Cliffs Nuclear Power Plant (NPP) near Washington, to NRC Region II in Atlanta, which gave a presentation of their regional activities, to the Institute of Nuclear Power Operations (also in Atlanta), to the Indian Point Nuclear Power Plants in New York, and to Brookhaven National Laboratory outside New York City.

Malyshev learned from top NRC officials that US plants were safer now than they were ten years ago at the time of the Three Mile Island accident, largely due to improved operator training, the widespread use of simulators, and increased management involvement in plant operations. Each side touched on the difficulties in bringing new reactors on line. As a result of the Armenian earthquake the US had sent a team of reactor safety experts to examine a Soviet nuclear reactor located in Armenia. Malyshev indicated that the Soviet public not only appreciated the assistance, but had also demanded that many of the improvements recommended by the Western experts in seismic safety be implemented within two years. He said that several other reactors in the Soviet Union, Hungary and Bulgaria would undergo some seismic redesign as a result of the December 1988 earthquake. In the end, both sides agreed on the importance of cooperation for improving reactor safety throughout the world.

The informality of his trip to NRC Regional HQ in Atlanta gave

Malyshev the opportunity to reflect on the problems nuclear energy regulators faced in the Soviet Union particularly after the accident. He noted that not only are most Soviet citizens ignorant about global nuclear power development, but the Soviet government faces such severe shortages that important commodities like fire resistant materials and necessary equipment for fire-fighting are unavailable for installation in Soviet reactors.[15] Malyshev reportedly left the Atlanta regional offices with a new appreciation for the applications of advanced technology to public safety – including on-line status reports of all US reactors, personal dosimetry, and emergency response capabilities – in the US, and particularly the decentralised power the NRC regional offices represented.

In Washington, the top Soviet regulatory official met with the NRC Executive Director for Operations and other key NRC officials in public affairs, regulatory functions, research, nuclear materials regulation and waste management, and operating data and emergency preparedness. He also signed the agreement for the exchange of inspectors, and seemed genuinely impressed with US openness as he commented on prospects for a mutually successful partnership.[16]

Following meetings with the New York Power Authority and with Consolidated Edison in New York, Malyshev conducted a press briefing. He was, he said, impressed by the extent of the NRC's authority and technical competence. He hoped to develop cooperation for establishing safety standards and to look into the possibility of acquiring equipment, particularly nuclear power plant simulators, to improve training of Soviet operators. And he said he hoped reactor safety could be enhanced through cooperation based on openness and trust. He admitted that prior to Chernobyl, the Soviets had placed too much confidence in the safety of their own nuclear plants. The accident led to a reorganisation of the Soviet regulatory apparatus, analysis of its legal basis, and enhanced personnel training; but more importantly, it had led to a change in attitude which now placed safety ahead of electricity production.[17]

In July 1989, as a follow-up to the Malyshev visit, two US NRC resident safety inspectors spent seven weeks in the Soviet Union learning all they could about a Soviet nuclear power plant.[18] Later that autumn, two Soviet inspectors took up similar duties at the Catawba nuclear plant in South Carolina. The purpose of this exchange under the JCCCNRS was to develop an understanding of the methods used by the Soviets to ensure adherence to design and operational requirements and specifications for their pressurised water reactors.[19]

The NRC inspectors reported 'exceptionally frank and candid' conversations with their Soviet counterparts and noted that the exchange was valuable in gaining insight into how the Soviet regulatory apparatus (GAEN) functions. Soviet inspectors, while quite similar to their NRC counterparts in the subject matter of their work,

> have much greater authority than NRC resident inspectors, as they can issue fines and stop work on their own authority. They are heavily involved in the line process of running the power plant and approving procedures and authorising maintenance activities. They are involved in post-event critiques and make recommendations for corrective actions.[20]

Differences increase up the chain of authority. In the Soviet Union, enforcement is based primarily on personal responsibility as opposed to corporate accountability in the US. As a result, while individuals are punished for errors, 'investigations conducted by GAEN into technical problems and operational events typically do not address programmatic weaknesses such as inadequate procedures, training, or management oversight'.[21]

At the local level, the NRC inspectors reported, the effectiveness of GAEN inspectors in resolving site-specific problems is diminished 'as identified problems become more systemic or generic, requiring corrective actions to be implemented from the regional or national level'.[22] Inspectors, and some of their superiors, were often frustrated regarding their inability to correct design weaknesses or influence the design agencies (reactor designers and production facilities in the USSR) regarding the need for greater attention to safety concerns.

The US inspectors' report added essential information on some of the generic weaknesses in the Soviet nuclear safety infrastructure that needed correction. The insights of the two US inspectors have proved extremely valuable in assessing the need for development of a 'safety culture' in the Soviet nuclear power programme, from shift operators and maintenance personnel all the way up through top management. They also put into perspective Soviet claims that their 'safety culture' has changed as a result of Chernobyl. Such deep attitudes, nourished over twenty-five years of reactor operation, cannot be changed overnight. The inspector exchange was therefore one of the most valuable, close-up looks the US has had into the way Soviet technical organisations operate. It demonstrated how far the Soviets must go to bring public accountability to their enterprises. But it also showed that both sides can learn lessons about how to improve their own functions by observing themselves in the mirror of their opposite numbers. It has been this opportunity for mutual benefit that is the reason for the success of the exchange programme. It may be that the Soviets do not see such a possibility for mutuality of interests in the area of health and environmental cooperation.

Health and environmental effects cooperation

In contrast to the remarkable progress made in reactor safety cooperation in the years after the accident, it was not until three and a half years after Chernobyl that the Soviets hosted the first exploratory meetings on environmental and health effects of Chernobyl under the JCCCNRS. These occurred in September 1989 in Kiev. While there were open and candid discussions on both subjects at this time and many of the Soviet participants appeared to offer opportunities for US participation in the analysis of health and environmental effects of the accident,[23] serious questions remained as to whether this cooperation will succeed.

The topics covered by Working Groups on (1) Environmental Transport of Radiation and (2) Health Effects were among the cornerstones of US interests in the cooperative exchange programmes. These subjects were expected to involve possible US participation in an integrated and long-term study of the medical impact of the Chernobyl accident, which could lead to a better understanding of the effects of atomic radiation on humans and on the ecosystem.[24] Instead, the Soviets turned to the International Atomic Energy Agency for such a study in October 1990.[25]

The situation in the Soviet Union presents a number of important areas for cooperation. At least three were considered to deserve priority attention:
1. a strategic plan for monitoring and treating children;
2. a study of the effects of moderate radiation on the thyroid;
3. a study of the increased incidence of leukaemia.

It was at these meetings that American participants fully came to understand the human costs of Chernobyl. However, they also became starkly aware of the depth of feelings and resultant divisions in Soviet society, and particularly the health community, over the problems of responding to Chernobyl's effects in a humane and effective manner. They learned that rainfall over the reactor accident plume led to large areas of contamination in the Ukraine, Byelorussia, and the Russian republic, with 'hot spots' occurring in areas hundreds of miles from the reactor, mainly in the direction of Moscow.[26] 'Protective actions',[27] such as evacuation, were based upon projected radiation doses of 10 rems in the first year. About 115,000 people were evacuated immediately, milk was discarded, and beef cattle were decontaminated or destroyed.

Other protective actions included liming the soil to drive down the acidity and plant uptake of radionuclides and the use of potassium rich fertilisers to block caesium uptake. Contamination of the surface water (at least initially) was not significant with respect to health protection. Radiation levels were, however, near the maximum permissible in several locations.

Soviet scientists said they had determined that children were the primary victims of Chernobyl, based largely on monitoring the uptake of Iodine-131 to their thyroids.[28] Six thousand children were detected as having received doses of over 200 rem to the thyroid. Fifteen children were born to women who suffered from acute radiation syndrome. The Soviets are now monitoring over 600,000 people (90,000 of whom are children) who received radiation doses as a result of the accident, and this is a massive undertaking. Some 300,000 persons were involved in the clean-up of the plant site itself, more than 100,000 have been permanently relocated and over 275,000 persons are still living in areas where 'rigorous radiation surveys' continue to be conducted. The 30-kilometre exclusion zone around the plant is still in effect.[29]

After studying close to 300,000 people and analysing the data, the Soviet authorities decided in 1988 that the maximum allowable individual lifetime (seventy years) dose for the population affected by the Chernobyl accident should be 35 rem (350 millisieverts or mSv). This limit included cumulative doses received from the day of the accident. The Soviets said they hoped to start to implement this limit as of January 1990.[30] About 20,000 people were reported to live in areas where the expected lifetime doses are in the range of 35 to 50 rem. Thus, anywhere from 20,000 to 100,000 may have to be relocated, under proposed intervention levels. Many members of the scientific community and the population, particularly in Byelorussia, have proposed even lower limits than 35 rem over a lifetime. Others, by contrast, want higher limits in order to get their farms back into production. This type of disagreement complicates matters for public health officials, and it is quite easy to see why many have been sceptical about the possibility of scientific cooperation with the West. What they want and need most is resources and assistance in dealing with the human problems.

US delegation leaders left the meetings convinced that no country can afford another accident of this magnitude. While they had known this on an intellectual level beforehand, their experience in Kiev brought the conclusion home with far greater force. Fortunately, the meeting produced several ideas for coordinated research. All the Soviet specialists with whom US team members spoke were in favour of cooperation with the US, but they were particularly interested in gaining technical support and additional resources for what urgently needs to be done. The Soviets clearly need help in gathering and interpreting the data using US techniques, instrumentation, and data processing. The help could take the form of training, provision of services or personnel to assist in laboratories. American specialists, on the other hand, are more interested in scientific cooperation to add to the body of scientific data for use by the rest of the world. They also believe that any cooperation must be based on full reciprocity and equal access to information, as if they were engaged in pure research. But the Soviets just as consistently assert that if they do not get the needed support, they do not have the time to do the various epidemiological studies foreign scientists would like to see done in order to further the knowledge of the scientific community.

The US team returned from Kiev determined to seize the opportunity to learn more about the real consequences of nuclear accidents, and about the effects of widely distributed but relatively low levels of ionising radiation, by supporting a major effort in cooperation with the Soviets in the health effects area. But they acknowledged that even an effort to study the three priority topics outlined above – and there seemed to be general agreement on these subjects – would take a huge cooperative effort. A number of proposals for cooperation in health and environmental effects were sent to the Soviets in January 1990, and the US side 'look[ed] forward to completing the comment and final selection process by March 30, 1990'.[31] The winter of 1989–1990, however, turned out to be one of 'discontent' for US officials who had been waiting more than three years to begin work on joint projects in health and environmental cooperation.

In a meeting in Moscow in January 1990, the US learned that the Soviets were still not ready to respond and appeared to lack coordination. While several US environmental projects had generated Soviet proposals, only one health effects response was presented. Even more surprising was a letter from the Soviet side which suggested a new area of cooperation for a radiation study and other areas for cooperation under a 'grant' or 'contract' arrangement.[32] Each of these carried a price tag for US involvement and were largely offering Soviet studies and data. They had not previously been discussed in relation to bilateral cooperation, and they were not real opportunities to share in the development and use of data from Chernobyl. Perestroika had hit the research community, and Soviet health scientists were apparently being asked to 'pay their own way' by their authorities.

The US requested that Working Group leaders on the health effects issue should get together in Moscow later that spring to reach agreement on *joint* research projects. Meetings were scheduled for late May, in Moscow, just after a high-level meeting of the US–Soviet Joint Committee (JCM) to work out details for renewing the US–Soviet Agreement on Peaceful Uses of Atomic Energy in Washington. The meeting of the JCM, in turn, was to occur the week before the Bush–Gorbachev Summit.

US planners had considered presenting the Soviets with an ultimatum at the JCM, which would have held up continued cooperation in other areas until agreement was reached on health effects, but decided on a more moderate course of action. The US complained that the Soviet response to their proposals had been a request for funds. One US health official noted pointedly to his Soviet counterpart in writing that:

We cannot fund grants. We can provide support for our people to travel and participate in projects in the USSR, to assist Soviet scientists who visit the U.S., and contribute to the means necessary to accomplish defined tasks. This leads to cooperative research, which means mutual planning of studies, review and sharing of data, and joint evolution of thoughts.[33]

Reporting from their Moscow meetings, however, the US delegation said there had been no change in the Soviet position on health-effects cooperation. Accordingly, the US co-chairman of the JCCCNRS complained about this lack of progress to the Soviet Minister of Nuclear Power and Industry, Vitaly Konovalov, at the JCM meeting in Washington which was also attended by a DOE Assistant Secretary, John Easton.[34] On Friday, 25 May at a meeting with Konovalov and Academician Evgeny Velikhov, Director of the Kurchatov Institute and Science Advisor to Chairman Gorbachev, Energy Secretary James Watkins 'again stressed the importance of the Health and Environmental Effects topic to the overall success of the JCCCNRS exchange'. Finally, a Joint Statement on Cooperation in Peaceful Uses of Atomic Energy was developed for inclusion in the Joint Statement of the May 1990 Summit between Presidents Bush and Gorbachev. This statement acknowledged specifically the agreement for joint cooperation in the study of environmental and health effects of past, present and future nuclear power generation.[35]

Three days later, as Chairman Gorbachev was leaving for the United States to seek US assistance in his economic crisis, the US Delegation in Moscow reported a dramatic change in Soviet attitudes towards health-effects cooperation. Perhaps the highest levels of the Soviet leadership had not previously understood the depth of the US interest in their information. In any case the judgement appears finally to have been that the overall interests of the Soviet Union in detailed health and environmental cooperation was one prerequisite to continued cooperation in other areas of nuclear safety. It remained to be seen whether this would be implemented at the working level.

Soviets request IAEA assistance

The difficulties in achieving US–Soviet agreement on health and environmental cooperation have probably been due as much to public controversy over the handling of the accident in the affected republics as to the requirements of perestroika. While the Soviet economic situation continues to deteriorate under perestroika, over a million people, including both victims and those trying to aid them, have had their lives and livelihoods disrupted by the Chernobyl accident. In timing that parallels the Kiev meeting of the JCCCNRS Working Groups on Health and Environmental matters, and in response to increasing levels of public criticism of their handling of the consequences of the accident, the Soviets asked the International Atomic Energy Agency (IAEA) in October 1989 to organise an international team of experts to investigate claims by local populations and local governments that large areas of the Soviet Ukraine and Byelorussia are unfit for continued human habitation.[36]

While Chernobyl resulted in only a few immediate fatalities and another 225 reported cases of acute radiation syndrome, the longer-term health consequences, especially on the children who were affected, are considerable. The Soviet social system, its legal and political institutions, and its public health facilities have been placed under intense strain. Publicity and public outrage and concern seemed to peak on the occasion of the fourth anniversary of the Chernobyl event, in April 1990. The availability of new facts and figures, the increasingly visible results of long-term effects of radioactivity, and particularly the burgeoning nationalism in Byelorussia and the Ukraine fed the fires. On 25 April 1990 a Supreme Soviet Press Conference held a lengthy discussion on Chernobyl. The first Soviet telethon was broadcast live from Moscow, with singing groups and interviews interspersed with footage of Chernobyl rescuers and children from the contaminated zones.

The Soviet press has reported protests by residents of the most contaminated areas, demanding better medical treatment, protection from radiation, and punishment for those involved in an alleged cover-up of the consequences of the accident.[37] The Soviet Chairman of the Emergency Committee of the Council of Ministers reportedly indicated that the USSR had spent 9.2 billion roubles on Chernobyl and that much of this has apparently been wasted: spent in the wrong ways and on the wrong things. Most of the future programme would consist of medical assistance. The central government plans to build four million square metres of housing, schools for 35,700 students, clinics able to accommodate 7,300 patients, and hospital space totalling 2,860 beds – all costing about 6.5 billion roubles.

During 1989–90, public opinion in the Soviet Union, encouraged by the reform movement in East Europe, became increasingly strident in its criticism of the Soviet nuclear power programme.[38] The Soviet co-chairman of the JCCCNRS wrote, for example:

Public opinion has become a new factor in this country, essentially affecting the energy policy. An incentive to wide public opposition to nuclear power was, of course, the Chernobyl accident. However, special attention paid to ecological problems as a constituent of the democratisation of the Soviet society has led to extension of public protest to a wide spectrum of energy facilities such as nuclear power plants and nuclear fuel cycle enterprises, hydroelectric plants, gas-production and coal-mining complexes, plants of fossil fuel processing, etc. Nevertheless, the 'synergetic' effect of the Chernobyl accident and the general ecological problems of the society have made nuclear power the most 'suffering' branch of the fuel-energy complex. There is practically not a single new NPP site where the local population would not protest against NPP construction . . .

Having no experience of formation of an unbiased public opinion, Soviet specialists who are convinced of the absence of any alternatives to nuclear power and at the same time approve a strict public control over the potentially dangerous modern technologies proved to be in a rather difficult situation.[39]

In response to the October 1989 Soviet request, the IAEA set up an international experts' 'Preparatory Mission' which visited the affected areas in March 1990 to identify major problems which should be included in an assessment by an international advisory team under IAEA auspices.[40] The visit gave the IAEA-led team an opportunity to observe the situation in the affected areas, listen to the concerns of the population, and begin to locate the type and amount of data that has been collected over the last four years. It became apparent to the participants that a vast amount of information has been collected. Unfortunately it is not all in one place, is mutually contradictory, and frequently is *ad hoc*.[41] Questions asked of the IAEA experts revealed the high levels of public anxiety about the health of their children from the numerous people that attended the meetings. People asked for the outside experts' views on the appropriateness of the 35 rem lifetime dose limitation, about the independence of the assessment team[42] and the future public availability of the assessment.

The final report is scheduled for completion by early 1991. Some US experts doubt that all the relevant data can even be organised by the team, much less analysed in an objective and comprehensive manner, and some fear a 'white-wash' imposed by the short time scheduled for the study. The Soviets appear to be seeking an independent judgement on their preventative actions following the accident in order to pacify an unhappy public. Should the IAEA report confirm that the Soviet Government acted properly in the health effects area, as the Soviets surely hope will be the case, the reputation for objectivity of the IAEA could be tainted.

Summary and conclusion

From the ambitious Protocol of the First Meeting of the Joint Coordinating Committee on Civilian Nuclear Reactor Safety in August 1988 to the present, implementation of the ten areas of cooperation has resulted in more than twenty-one substantive working group meetings and exchanges. Scores of technical documents have been exchanged. More than seventy-five Soviets have attended meetings in the United States, and at least an equal number of American officials have travelled to the Soviet Union under the auspices of this programme. A major inspector exchange has taken place which sent three Americans (two reactor inspectors and one translator) to the Zaporozhye nuclear power plant in the Soviet Ukraine and brought three Soviet technicians to the Catawba nuclear power plant in South Carolina. A Second Protocol of the JCCCNRS, signed on 31 October 1989, outlined areas of cooperation and work to be done in 1990 and beyond and called for no less than twenty-five working group meetings in the two countries during the fiscal year 1990 (October 1989 to September 1990).[43]

From the outset, the US recognised that it would be a long and difficult process to understand Soviet designs and approaches to nuclear safety, much less to influence these approaches, especially in view of the isolation of the Soviets from interaction with the West for many years. Accordingly, a multi-year effort was begun, with effort concentrated on the ten most important areas.[44] The United States has been careful to avoid situations in which the NRC and the US Government could be construed as giving a seal of approval to the Soviet nuclear programme. In its cooperative activities, the NRC has provided tools the Soviets can use within the constraints of its own system to enhance safety. The approach has been a person to person approach. The Working Groups offer a chance for experts to get to know one another well enough to trust the other's judgement, much as the original exchanges in 1986 were born in the non-confrontational approach to the Soviet technicians in Vienna. Through these professional ties, each side has shared its own experience and discussed mutual problems. The US continues to avoid placing the Soviets on the defensive while trying to build up mutual confidence in the benefits of cooperation. Accordingly, substantive exchanges have taken place and positive attitude changes have resulted on both sides. The US and others hope this may result in future changes in safety 'philosophy' in the Soviet nuclear bureaucracy.

But it is evident from strained cooperative ties in the health and environmental areas that all is not working as US planners wish. The sources of these strains are probably the result of both perestroika, which has required individual institutions to make their own way and acquire their own resources, and glasnost, which has permitted the politicisation of the Chernobyl health-effects debate in the USSR. It took preparations for a Bush–Gorbachev Summit to break the log-jam which finally resulted in bilateral cooperation. And the request for the IAEA study was motivated by the desire to mollify public criticism.

The Chernobyl accident was one of the most important factors in spurring the Soviet political crisis following the implementation of the twin policies of glasnost and perestroika.[45] Chernobyl came at a time in Soviet history, with the ascendance of Gorbachev and the deterioration of the Soviet economy, when its international impact forced the Soviet régime further along a path already charted but at a pace that made it impossible to slow down, particularly in a country as weakened as the Soviet Union. Chernobyl was the event that catalysed public environmental concern, indeed legitimised it, and forced decision makers to look beyond their own borders for broader solutions. The United States took an appropriate stance in response to the Chernobyl crisis, a stance which has paid rich dividends in furthering technical cooperation between the two countries. Only a return to pre-liberalisation policies of rigidity and centralised control can disrupt the growing nuclear safety régime that is being created both bilaterally with the US and multilaterally in the IAEA.

Notes

The views in this paper are those of the author and do not reflect the views of either the US Nuclear Regulatory Commission or the IAEA.

1. The protocol outlines a programme of cooperation in ten technical areas that has been implemented in a series of Working Group meetings over the last two years. The ten technical areas are listed in note 12.
2. See: Edwin Morse, *Modernization and the Transformation of International Politics* (New York and London: The Free Press, 1976).
3. The concept of 'régimes' is spelled out in Joseph S. Nye and Robert O.

Keohane, *Power and Interdependence* (Boston: Little, Brown, 1977). Cf. Seyom Brown and Larry L. Fabian, 'Towards mutual accountability in the nonterrestrial realms', *International Organization*, 29, no. 3 (1975), pp. 877–92.
4. An agreement signed at the height of détente in 1973 by President Nixon and Chairman Brezhnev but which had lain dormant for several years following the Soviet invasion of Afghanistan.
5. These were the three original topics for cooperation in the 1973 agreement.
6. These items were later to become the basis for more specific agreements on safety exchanges and cooperation. From Francis Gavigan and R. W. Barber, US Department of Energy, Report of a Trip to the USSR, 'Report of a trip to the USSR by a US Government Nuclear Safety Team, March 2–13, 1987', unpublished trip report.
7. Much of this information, taken from the Gavigan/Barber trip report, has later turned out not to be true. Subsequent study, according to the Soviets, showed no differences between mental retardation in births within exposed population and births elsewhere. This the Soviets attributed to rapid evacuation of the 1,000 pregnant women exposed throughout the Ukraine (about 200 of whom were in the critical period of radiation sensitivity). None received more than a 0.5 rem radiation dose. The Soviets also indicated that no reduction in head size was noted in those born after Chernobyl – considered a good indication of normal mental ability.
8. See: *Extended Synopses, International Conference on Nuclear Power Performance and Safety*, Vienna, Austria, International Atomic Energy Agency (IAEA-CN-48), 28 September–2 October, 1987.
9. As a measure of the significance of this reciprocal Soviet trip to the US, just a month and a half after the visit the Joint US–Soviet Summit statement issued by President Reagan and General Secretary Gorbachev on 10 December 1987, at the conclusion of Gorbachev's first visit to the US, included the following reference to nuclear safety cooperation: 'The two leaders noted with satisfaction progress under the bilateral agreement on peaceful uses of Atomic Energy towards establishing a permanent working group in the field of nuclear reactor safety, and expressed their readiness to develop further cooperation in this area'.
10. See: DOE NEWS, 'US Energy Secretary Herrington releases interagency guidelines for nuclear safety assistance to Eastern Europe', 2 July 1987, which provided the framework for a coordinated US approach to this facet of cooperation. Herrington's statement said: 'We have developed a set of guidelines which encourage US Government and industry cooperation and coordination in providing assistance in nuclear safety-related activities in Eastern Europe and the Soviet Union. Through broad distribution of the guidelines to industry, we hope to maximize the commercial potential for nuclear safety-related sales and thus achieve an enhanced degree of power reactor safety'.
11. See: United States Nuclear Regulatory Commission, 'News Releases', week ending September 14, 1988, NUREG/BR-0032, vol. 8, no. 35, for text of the Protocol.
12. 1. Safety Approaches and Regulatory Practices; 2. Analysis of the Safety of

Nuclear Power Plants in the USSR and the US; 3. Radiation Embrittlement of the Housing and Support Structures and Annealing of the Housings; 4. Fire Safety; 5. Modernisation/Backfitting; 6. Severe Accidents; 7. Health Effects and Environmental Protection Considerations; 8. Exchange of Operational Experience; 9. Diagnostics, Analysis, Equipment, and Systems for Supporting Operators; 10. Erosion/Corrosion, Destruction of Piping and Components.
13. For a more thorough description of this visit, see: 'Report of visit of the Soviet delegation headed by Chairman Vadim Malyshev, May 16–24, 1989', US NRC Public Document Room, SECY 89–252, 18 August 1989.
14. Malyshev's concerns about decommissioning are a result of the small reserves of electrical generating capacity in the USSR. He said a great deal of attention is being placed on plant life extension, since shutting down reactors will cut into this reserve, although the impending shortfall in electrical production is to be made up in the immediate future through gas turbine generation. Ironically, this is also what is happening in the United States, as few utilities are willing to invest large amounts in huge electricity generating plants, especially nuclear plants, in the current uncertain economic and energy demand market.
15. Trip Report by Dr Ed Shomaker, US NRC, 30 June 1989. Available in the Public Document Room, US Nuclear Regulatory Commission, Washington, DC.
16. This presentation was so favourable that Malyshev, upon his return, sought to fashion his own regional regulatory structure after the decentralised US model. See: Ann MacLachlan, 'Soviets look to NRC as model for revised nuclear regulatory body', *Inside NRC*, Washington, DC, 11, no. 21, 9 October 1989, pp. 1 and 14–15.
17. Shortly after his return to the Soviet Union, Malyshev's US visit took on added significance when he was promoted by the Supreme Soviet to be chairman of a newly created safety regulatory body (the 'GPAN') which covers all industrial safety matters including nuclear. In conversations with US officials almost a year later, in March 1990, Malyshev said that his greatest challenge was to define an appropriate role for the republics in the supervision of nuclear energy. He said he hoped to adopt for the USSR the US regulatory model and asked for copies of agreements the NRC has with the local governments of states such as Florida and Illinois, under which the central regulatory body, the NRC, grants regulatory rights to these states for supervision of medical and industrial applications of nuclear materials.
18. L. J. Callan, Director, Division of Reactor Safety, NRC Region IV, and P. B. Brochman, Senior Resident Inspector, Clinton nuclear power plant, NRC Region III, travelled to the USSR from 8 July to 27 August to serve at the Zaporozhye nuclear power plant (ZNPP) under the regional office (Kiev) of the USSR State Committee for Supervision of the Safety of Work in Industry and Atomic Energy (GosAtomEnergoNadzor or GAEN) of which V. Malyshev had been director.
19. The Trip Report prepared by Callan and Brochman is dated 14 November 1989, and is available in the NRC Public Document Room. Extensive excerpts of their report on operations at the Zaporozhye plant are in 'NRC

urged to study Soviets' approach to radiation control at Zaporozhye', *Inside NRC*, (a Nuclear Industry Newsletter), McGraw Hill, vol. 12, nos. 9 and 10, 23 April, p. 3, and 7 May 1990, pp. 5–7.
20. Ibid.
21. Ibid.
22. Ibid.
23. NRC Public Document Room, SECY-89-351, 'Update on US–USSR civilian nuclear reactor safety cooperation', 24 November 1989.
24. Ibid., Enclosure 1, p. 1. See also November 1989 trip reports by R. Martin and S. Yaniv, available in the US NRC Public Document Room. It is clear that the data on health effects from Chernobyl are far more 'realistic' for the study of nuclear power plant accidents than those from the only other large-scale radiation incident – the bombing of Hiroshima and Nagasaki – which contributed the bulk of data on which radiation dose limits are now based. Obviously, the bombs caused instantaneous doses quite unlike the more probable forms of low- to medium-level doses being seen in the Soviet Union. The Soviets also make this point in justifying a Chernobyl Research Centre to be sponsored in part by the International Atomic Energy Agency.
25. See later section headed 'Soviets request IAEA assistance'. These effects could also be crucial to the future development and use of nuclear power as an energy source.
26. What was not learned until later, however, is that maps showing this wide spread of contamination were not made public until March 1989, almost three years after the accident. The information was undoubtedly explosive, and keeping this data secret no doubt contributed heavily to the public outcry in the Soviet Union over the lack of information made available to the public by the Soviet authorities.
27. 'Protective Actions' and 'Preventative Measures' are terms that refer to means authorities use either to limit or to mitigate the effects of radiation releases. One example is administering iodine tablets to exposed populations to saturate their thyroid with benign forms of iodine, to prevent radioactive iodine from being deposited in the thyroid, which is the body's repository for this element. The most serious protective action, of course, is evacuation.
28. 'Our children suffered the worst of it', one US participant reports hearing from Academician L. A. Ilyin of the Institute of Biophysics of the USSR Academy of Medical Sciences at the Kiev meeting in September 1989. Thyroid exposures of children were most threatening, with children drinking caesium-contaminated milk for a month after the event, and contamination has still not disappeared. Caesium has a 30-year half life and is deposited in the muscle. *Trip Report*, Robert W. Miller, US Nuclear Regulatory Commission, 13 October 1989.
29. The most authoritative account of this meeting is the 'US–USSR Joint Coordinating Committee for Civilian Nuclear Reactor Safety (JCCCNRS) Memorandum of Meetings: Working Groups 7.1 and 7.2', issued about 30 October 1989. The document is available in the NRC Public Document Room. It concludes:

The main thrust of these meetings was a candid exchange of problems on both sides. The Soviets indicated that they have a unique fund of data in relation to the Chernobyl accident and that many other nations have approached them to participate in various forms of cooperation. They are concerned that whatever research of cooperative activities are carried out must be accurate, must add credibility to all participants, and must live up to public scrutiny. The US side completely agreed with these points.

30. Because of opposition from officials in the Ukraine and Byelorussia, however, this strategy was amended. The final implementation may not take place until 1992. For a detailed discussion of this issue, see the forthcoming *Report on the Health Effects of Chernobyl*, due to be published by the IAEA in mid-1990.
31. These were transmitted in a letter from NRC Executive Director for Operations, James Taylor to his Soviet co-Chairman, N. N. Ponomarev-Stepnoy, dated 19 January 1990, available in the NRC Public Document Room.
32. In March 1990 the project coordinator for the US-Soviet cooperative programme met with Dr L. Bolshov, First Deputy Director of the Institute for Nuclear Safety, which is sponsored by the USSR National Academy of Science. The institute is under the directorship of Dr E. Velikov, who also heads the Kurchatov Institute. Dr Bolshov was asked about the possibility of success in Working Group 7. He replied that Dr Ilyin is under some criticism for certain public statements made shortly after the Chernobyl accident. The ministry had also delayed certain evacuations of people from contaminated areas, apparently to play down the radiation risk, and authorised articles that tended to dismiss serious radiation problems. See: Trip Report of Dr Edward Shomaker, 'Visit to the USSR, March 5-17, 1990', 18 April 1990, available in the NRC Public Document Room. An example of such an article cited was given to Shomaker, *viz.*, 'Ecological features and biomedical consequences of the accident at Chernobyl NPP' (translated from Russian), *Radiation Medicine Magazine*, March 1989.
33. Ibid.
34. US co-chairman, Taylor, wrote that 'I stressed the need for Soviet leadership to show progress on the topic of Chernobyl Environmental and Health Effects to the overall success of the JCCCNRS exchange. Assistant Secretary Easton reiterated the importance of this topic in his summary remarks at the end of our meeting'. See: 4 June 1990 Memo from Taylor to the NRC Commissioners, topic: 'Ninth US-USSR Joint Committee Meeting', NRC Public Document Room.
35. Ibid. The text of the Joint Statement is included as Enclosure 2.
36. The subject was first broached in September 1989 by Alexander Protsenko, Soviet Governor to the International Atomic Energy Agency, in Vienna at the General Conference of that body. Cf. 'Chernobyl's radiological consequences', *IAEA Newsbriefs*, 5, no. 4 (44) (May 1990).
37. The Soviet request for IAEA assistance came in the wake of 'a rash of new and alarming assertions by Soviet officials, inspired to discovery by glasnost. They are now saying that the scope and damaging health and environmental effects of the accident were covered up by "Stalinists" in the Soviet

bureaucracy and that the impact of Chernobyl is catastrophically far worse than previously known'. *Nucleonics Week*, 31, no. 19 (10 May 1990), pp. 3–4. An example of such a report, albeit unvalidated, was an assertion by the founder of the Soviet Green World movement that radioactive contamination in the Kiev reservoir along the Dnieper River could eventually force the evacuation of Kiev. 'Contaminated reservoir may force Kiev evacuation, says ecologist', *Nucleonics Week*, 31, no. 19 (10 May 1990), pp. 5–6.

38. Jane I. Dawson, 'The emergency of the anti-nuclear power movement in the USSR', unpublished paper delivered at the Conference on Psychological Aspects of the Chernobyl Accident, Gomel, Byelorussia, USSR, October 1990.

39. N. N. Ponomarev-Stepnoy, and A. Yu. Gagarinskii, 'Nuclear power in the USSR: status, prospects, public opinion', unpublished paper presented at the Seventh Pacific Basin Nuclear Conference, San Diego, California, 4–8 March 1990, pp. 35–6.

40. On 7 May 1990, the International Atomic Energy Agency announced the issuance of a draft report entitled 'The radiological consequences in the USSR from the Chernobyl accident: assessment of health and environmental effects and evaluation of protective measures'. This report summarises the Soviet initiative and outlines Agency actions either taken, or to be taken, in implementing the international programme to perform an independent inspection of several sites in the Soviet Union. Information in the following paragraphs is taken from this report. The report is also referred to in *Nucleonics Week*, 31, no. 19 (10 May 1990), pp. 3–4.

41. The team was approached by various local groups which were willing to provide 'unofficial data' for use in the assessment. This is similar to certain low-level health officials who have sought to provide data to the United States Government, some through private firms representing them in Washington. Any scientific programme, while not ignoring such offers, must view them with a healthy scepticism, since those offering the information usually have a cooperative programme in mind requiring Western resources.

42. In a number of towns visited by previous international missions, no changes had resulted from discussions with local citizens, leading to scepticism about the bona fides of the IAEA team.

43. One small measure of progress is indicated in 'Soviets look to NRC as model for revised nuclear regulatory body', *Inside NRC*, 11, no. 21 (9 October 1989), pp. 1, 14ff, which notes, 'If the plan works out, in a few years the USSR nuclear power industry will be going through three sets of hearings to site, build, and operate nuclear plants, with public hearings at all steps of the way, and regional offices and resident inspectors will monitor the safety of reactors far from Moscow'.

44. Areas in which the United States has taken the lead are regulatory law and activities, backfitting, Probabilistic Risk Assessment (PRA), simulator and computer training, design reviews of safety systems, and severe accident analysis. The USSR has unique experience and data in annealing and embrittlement, pump seals and, of course, the environmental and health effects of Chernobyl.

45. See, for example, Zhores Medvedev, *The Legacy of Chernobyl* (New York and London: W.W. Norton & Co., 1990), who argues that in the four intervening years the ill wind blowing from Chernobyl became the principal factor in the Soviet Union's political crisis. He quotes V. Legasov, who compared the historical significance of Chernobyl to the eruption of Mount Vesuvius which buried Pompeii in hot ashes in AD 79. This book surveys a previously unpublished picture of Chernobyl's impact on Soviet energy supplies, agriculture and the environment, calculating the costs in billions of dollars.

11 The global impact of the Chernobyl accident five years after

Zhores A. Medvedev
(Translated by Åse Berit Grødeland)

For the first days, weeks, and even months after the Chernobyl accident, those nuclear powers which not only make use of nuclear energy but also manufacture reactor equipment and construct nuclear power plants on their own territory as well as in other countries were concerned about the impact of the global fall-out of radionuclides. First and foremost, though, they tried to convince public opinion that other reactor types could not be subject to a similar accident.

In 1986 there were 428 energy-generating nuclear reactors in operation in 25 countries. Another 149 reactors were either under construction or were being planned. Nuclear energy already provided 16 per cent of the world's electricity production. In some countries, however, the relative share of nuclear power plants in the total energy balance was far higher. The fate of nuclear power depended upon an exact analysis of the causes of the accident.

Prior to the Chernobyl accident only one 'ultimate' accident (a core meltdown) had occurred in the nuclear power industry: the meltdown of the uranium fuel in the upper part of the reactor at Three Mile Island in Pennsylvania on 28 March 1979. But this was only an 'economic' catastrophe. No human lives were lost and the emission of radioactivity into the environment was limited to inert radioactive 'noble' gases with a short half-life, thus not constituting any serious danger. Usually these gases are discharged in one way or another into the environment by all nuclear power plants, as there is no means by which to absorb them.

According to calculations made by experts projecting and constructing nuclear power plants in the United States, one ultimate accident resulting in the meltdown of radioactive fuel and a reactor loss could happen in the course of approximately one million reactor years, that is once every thousand years per thousand reactors. These calculations constituted the basis for the classification of the production of nuclear-generated electricity as a 'relatively safe technology'. Despite this, the major means of shielding the reactors – which included the construction of very solid domes of reinforced concrete to cover them – were intended to protect the population even from an ultimate accident. The fact that the Three Mile Island accident took place not in the middle, but at the very beginning, of the nuclear era came as a big shock to the nuclear industry. The US Department of Energy conducted a thorough analysis of the causes of the accident, resulting in the introduction of more than 100 new construction and safety requirements to safeguard nuclear power plants which considerably increased construction costs.

Prior to the accident in 1979 all estimates of prices for electricity produced at nuclear power plants indicated that they had great advantages compared to electric power stations running on fossil fuels. After 1979 these advantages were lost. Moreover, the production of electricity at nuclear power plants was reclassified from being 'relatively safe' to 'potentially dangerous'. From 1980 onwards, the likelihood for a similar accident occurring was increased to one every 20,000 reactor years, in other words increased 500 times.

The Chernobyl accident caused such a shock among Western experts and atomic specialists, not only because even this prognosis turned out to be over-optimistic, but also because it was a catastrophe that was not considered 'possible in principle' and was not even envisaged by the reactor safeguarding systems. This was not a simple core meltdown due to the heat from residual fission radionuclides generated in the reactor as a by-product of the controlled chain reaction of uranium-235, the main nuclear fuel. It had occurred instead as the result of a runaway chain reaction of uranium-235. In technical terms, this phenomenon was referred to as 'reactivity excursion'.

The accident that took place in the USA on 28 March 1979 was facilitated by the meltdown of radioactive fuel, which developed for 2 hours and 54 minutes. At Chernobyl, the reactivity excursion continued only for a few seconds and resulted in the complete destruction of the reactor and the reactor building. Against a reactor explosion of this kind, any safety system would prove inefficient. The fate of nuclear electric power now rested upon an accurate analysis of the causes and the nature of the explosion.

The accumulation in the reactor's uranium fuel load of numerous short- and long-lived radionuclides cannot exceed the fuel's initial content of the isotope uranium-235. In the case of the Chernobyl reactor, operating on uranium oxides (a mixture of isotopes 238 and 235 containing uranium-235, enriched to 2 per cent) the total contribution of radionuclides releasing heat in the reactor by the end of the two-year reactor cycle similarly cannot exceed 2 per cent. In a pressurised water-moderated reactor (PWR), with more enriched fuel containing 4 per cent of isotope-235, the accumulation of radionuclides at the end of the reactor cycle is higher. In any event, however, the heat that may cause the reactor-fuel to melt down if the coolant is lost and the emergency accident cooling fails to work (as happened in the United States in 1979) constitutes at the most 4 per cent of the total heat released by the reactor when operating normally at full power.

At Three Mile Island the accident took place during the initial stage of the nuclear cycle, four months after the reactor had been started up in late December 1978. In this case the output of heat from the fission radionuclides did not exceed 1 per cent of the reactor power.

As a result of the sharp temperature increase, hydrogen was released when the reactor fuel melted down. This caused the danger of a combustion. The hydrogen combustion which took place 9 hours and 30 minutes later, however, did not damage the protective containment cap. A light gaseous hydrogen combustion is not very powerful as, in contrast to when gunpowder explodes, no gas is created. On the contrary, it disappears when water is being formed.

According to estimates, the explosion did not exceed 10–15 kg TNT – comparable to the power of a small shell or an anti-tank mine. The reactor's protective cap, judging from assertions made by the engineers, could withstand terrorist bombs and the weight of an airliner crashing onto it from above.

According to information presented by the USSR at the post-accident review meeting of the International Atomic Energy Agency (IAEA) in Vienna at the end of August 1986, the unexpected runaway chain reaction of uranium-235 in the Chernobyl reactor, i.e. the so-called 'reactivity excursion', exceeded the projected reactor power by a factor of 100 prior to the moment when the first explosion was registered by outside monitors.

The explosion pushed away the upper protective plate – weighing more than 1,000 tons – and threw parts of the reactor-fuel upwards. The intensity of the local heat release during this reactivity accident exceeded the Three Mile Island accident parameters 10,000-fold. No adequate protection against this kind of 'reactivity excursion' existed. If the reactor at Chernobyl had been placed not inside a thin steel cylinder with the upper plate passively resting on rollers but placed inside a hermeticised steel vessel, as is the case in water-moderated models in the United States and other countries, then the runaway chain reaction could have continued with impacts that would be difficult to predict. The explosion could have been far more powerful and could have blown everything not only up into the air, but also in a horizontal direction. Close by there were another three reactors, operating at full power.

The Western reconstruction of the accident

At the international experts' meeting at the IAEA in Vienna on 25–29 August 1986 devoted to the analysis of the Chernobyl accident, the Soviet side presented a detailed report including extensive annexes. The report's major conclusions were published in the Soviet journal *Atomnaya Energiya* in November of the same year. The IAEA also published a short account, based on the report and the debate, as early as a month after the Vienna meeting. These materials had a markedly preliminary character and so far remain the main technical documents concerning the development of the Chernobyl accident. However, the other major nuclear power, the United States, was not fully satisfied with the Soviet documents, as they left too many questions unresolved.

The British delegation to Vienna, headed by Lord Marshall, Chairman of the Central Electricity Generating Board, which was also responsible for all nuclear power plants in Great Britain, was not so sceptical and accepted the Soviet report almost in full. In the USA all nuclear power plants are private joint stock property. The Federal US Department of Energy only controls their activities, including safety problems. Therefore American experts from this department were most interested in understanding what had happened at Chernobyl and, based on this, in elaborating new safety rules. Besides, new reactors were no longer being constructed in the United States in 1986. No orders for the construction of nuclear power plants had been placed since 1979. In Great Britain nuclear power plants are state-owned, and a plan existed to double the generation of nuclear energy by the year 2000. This accounts for the differing approaches.

France took no active part in the debates at all, as it depended heavily on nuclear energy – receiving 70 per cent of its electricity from nuclear power plants. Japan, half way through its very energetic nuclear power plant construction programme, did not send a single expert to the IAEA Vienna meeting. The Japanese 'Nuclear Forum' obviously did not wish to take into account what had happened at Chernobyl, holding the view that this was of no relevance to their far more advanced technology.

The American delegation arrived in Vienna earlier than the others and in the USA all the documents sent from the USSR in Russian only were quickly translated into English independently of the IAEA. During the meeting all illustrations and slides that were shown by Soviet speakers were copied and qualitatively improved, 'enriched' with the help of computers. The two-volume work *Soviet Illustrations presented at the IAEA Meeting in Vienna* was distributed among all interested organisations. Having returned home, the American experts immediately started testing all the Soviet models of the accident on their supercomputers. The Soviet computer model of the accident presented in Vienna did not satisfy the Americans. Missing explanations and mistakes were detected in it. The computers, of the 'Scala' type, that had served reactors of the Chernobyl type and were run on a special programme called 'Prisma', were not sufficiently up to date. The Americans wanted to test all the information on their own computers and with their own programmes.

The major uncertainty remaining after the Vienna meeting was cardinal: the cause of the explosion. Apart from the many mistakes made by the operators, the Soviet expert group, headed in Vienna by Academician V. Legasov, admitted only one design fault, termed 'the positive reactivity void coefficient'. Graphite was the main moderator of neutrons in the reactor, making it possible to maintain the chain reaction of uranium-235 at a specific level.

The fission of uranium-235 releases both fast and slow neutrons. It is very difficult to control the fast neutrons (which can be done only in certain 'fast' reactors) and therefore they are delayed by some substances with light atomic nuclei. Graphite has been considered a suitable moderator of neutrons since the first reactors were built in the United States in 1943. However, the water that cools down the reactor is also a good means by which to slow down the neutrons, as it contains light hydrogen atoms. The appearance of a large number of steam bubbles in the water, circulating in the reactor channels, reduces its ability to slow down the neutrons. As a result, the power of the fast neutrons' field increases, which in turn speeds up the heating process and consequently further increases the production of steam. If evaporation becomes too extensive in the reactor core, a runaway nuclear reaction on fast 'prompt' neutrons may start. To prevent this, control rods containing boron – powerful absorbers of neutrons – are quickly inserted into the reactor core, thus shutting it down.

Judging from all the documents that have been presented by the Soviet side, precisely this construction defect, 'the positive steam void coefficient', caused the accident. Due to mistakes made by the operators, who withdrew from the core almost all the control rods and infringed other safety instructions, the evaporation of water in the reactor intensified and consequently reactivity, i.e. the release of heat, quickly increased. This in turn caused the runaway as the increased release of heat also increased the volume of steam, displacing the water and reducing the efficiency of the cooling. The operator decided to shut the reactor down, but it takes too long – 20 seconds – to lower the control rods in this model. This was enough to cause the first explosion –

The global impact of Chernobyl five years after 179

4 seconds after the scram button to shut down the reactor had been pushed.

The cause of the explosion itself, which blew the upper steel plate weighing more than 1,000 tons into a vertical position and tore into pieces the thousands of steel pipes running through this plate, was in the Soviet assessment to the IAEA explained by a mixture of gases – hydrogen and carbon monoxides – which formed as a result of the interaction of overheated steam with zirconium and graphite. The formation of such gases when the local temperature increased to more than 2,000°C could indeed take place. When mixed with oxygen from the air, an explosion should occur. It has been assumed by various commentators before and after the IAEA meeting that such a conventional 'thermal' explosion caused the accident. In official Soviet literature the assertion that the explosion was not 'nuclear', but conventional, is still repeated.[1]

American experts did not agree with this explanation. Firstly, a gas explosion could not push up such a heavy plate having first torn apart an enormous number (more than 1,000) of strong steel pipes. Secondly, chemical thermal reactions do not take place so instantly and simply could not accumulate enough gases in the course of 3–4 seconds to cause a powerful explosion. Besides, the description of the accident included two explosions, the second following almost immediately after the first.

During the discussion in Vienna on 25–29 August 1986, the first Soviet version of the accident was rejected. In the IAEA's concluding document, distributed on 24 September 1986, the first explosion that lifted up the upper plate of the reactor 'weighing more than 1,000 tons' is explained by a sharp increase in pressure inside the reactor caused by increased reactivity and the extensive release of energy, i.e. in reality a chain nuclear reaction. As for the second explosion, which followed 2–3 seconds after the first, the IAEA report is less definite, noting that so far it is not clear whether the explosion was caused by hydrogen being formed and reacting with air, or whether it was caused by a second burst of power. The answers to the questions that were raised required additional analysis. More than 600 technical questions were handed over to the Soviet side for further explanation.

However, the testing of all the Soviet information about the course of the accident on American computers run on various programmes indicated that the accident had a more complex physical character. As early as November 1986 the US Department of Energy distributed among interested organisations its detailed and independent assessment of the accident.[2] The American experts' major 'discovery' was the disclosure of the fact that a certain increase in the reactor's reactivity in its lower part began immediately after the scram button had been pressed. The upper part of the reactor was inhibited. In the lower part, however, the reactivity increased. That in itself indicated construction defects in the control rods themselves.

Diagrams of these rods received from Soviet colleagues at the request of the IAEA – they are included as illustrations in the American report but were missing in the Soviet account at the IAEA – provided the answer to the confusion. The protection control rods did indeed have a construction defect: the graphite water displacers were too short. When the control rods were lifted to the upper position, the graphite water displacers proved to be in the middle of the core, 125 cm from the bottom of the reactor core.

Graphite is an active moderator of neutrons and obviously for this reason a decision was made to add the water displacers to the 1,700-ton graphite blocks that formed a component of the core. When the rods are constructed in such a way that the graphite water-displacer in their channels occupies only the central part of the reactor, pressing the scram button causes the lowering of the water displacers. Meanwhile, the speed of the chain nuclear reaction in the lower part of the reactor accelerates in the course of the four seconds it takes for the graphite displacer to reach the reactor bottom. But even after this the reactivity may increase in the course of the sixteen seconds before the boron-containing neutron absorber reaches the lower part of the reactor. Four seconds is a long period of time for nuclear fission processes. After 3.9 seconds, according to the results of the American analysis, the reactor's power steeply increased by 109-fold! This resulted in the first explosion.

The American assessment identified this defect of the control rods as the 'positive scram effect', i.e. 'runaway instead of a stop'. The description of this defect as the major cause of the accident was included only in the second edition of the British assessment of Chernobyl in April 1988, after Soviet experts had confirmed this analysis. The first person to address the defect in the Soviet Union was Grigory Medvedev, in his documentary novel *Chernobyl Notes*, written in 1987 but not published until 1989.[3] However, in January 1988 the group of Soviet experts who had published the analysis of the first stage of the accident in the journal *Atomnaya Energiya* (no. 1, 1988) also acknowledged that the construction defect of the control rods was the initial cause of the accident. The authors conducted a new computer analysis of all the parameters and admitted that after the emergency protective system of the AZ-5 had been activated, 'the increase of neutron power in the lower part of the reactor after 1.5 seconds starts to prevail over the reduction of neutron power in the upper part, as the average integrative power increases, reaching a maximum at 7.5 seconds, after which it starts dropping quickly'. In these circumstances, 'in some energy-strained parts of the core the temperature of the fuel elements exceeds the melting temperature of the fuel'. This must cause the stem to overheat, and 'destroy the technological channels and the entire reactor'.[4]

It was this destruction of the lower part of the reactor core which stopped the control rods from moving and caused the explosion. The American assessment also clarified the causes of the second explosion, having rejected the possibility of its occurrence as a result of the accumulation of hydrogen. The first explosion was an obvious result of the runaway chain reaction caused by the graphite displacers being pushed downwards and by the positive steam void coefficient. But the second explosion, as shown by the computer analysis, took place only one second after the first one as a result of the second massive increase in power, 470 times greater than the reactor's projected maximum power. At this moment an enormous amount of thermal energy was released in the reactor – enough to melt down the reactor fuel and increase the pressure of various fumes, steam and gases to levels that could displace the upper plate and everything that had been mounted on it into the reactor hall, throwing up nuclear fuel, pieces of graphite and fragmented parts of the fuel channels.

These conclusions were accepted by Soviet experts during subsequent contacts with foreign colleagues at the IAEA. Consequently they were included in the second, revised edition (for official use) of the British assessment of Chernobyl compiled by fourteen authors. However, the second increase of reactivity was calculated, in agreement with Soviet physicists, as 'approximately 440 times the full power' which the temperature increased to 3,150°C. In this way the flaw in the design of the protection control rods was identified as the major cause of the accident as early as 1986. To the British and the Americans it was absolutely incomprehensible how such an elementary design flaw could have been permitted. The assessments stressed that neither in the United States nor in Great Britain would it have been possible for any construction flaw to depreciate the protective system to such an extent. The Soviet side was unable to come up with any sensible explanation as to why the graphite water displacers did not cover the lower part of the reactor. It was known (it had been confirmed after the publication of the article by Adamov and others in *Nuclear Energy* in January 1988) that after the Chernobyl accident all the other reactors of this type (RBMK) had been fitted with control rods that were partly lowered so that the water displacers covered the entire lower part of the reactor. But in this case the reactor failed to operate on full power as its upper part was inhibited. Therefore, as a result, a simpler solution was found – the part of the emergency protection control rod that links it to the graphite water displacers was simply enlarged from 125 to 250 cm.

In the preface to the second edition of the British assessment, Lord Marshall very strongly criticised this construction fault:

However, in addition to the design defects that were identified in our first report, it now seems likely that the magnitude of the accident was enhanced by a further phenomenon called 'positive scram'. This effect was first identified by a US Department of Energy team and has now been confirmed by the Russians. It is described in later sections of this report. It seems that in some unusual configurations the control rods of a Chernobyl-type reactor do not always act to reduce and control the fission process. In circumstances when they are first activated the initial effect is to *increase the reactivity instead of decreasing it*. Such a situation would have been unacceptable to the designers of a Western reactor but the Russian designers thought they could live with this strange phenomenon because of the low probability that this particular configuration of control rods would come about. But, as it happens, it did come about, it did come about in the Chernobyl incident, and it now appears that this phenomenon enhanced the magnitude of the accident.

The British assessment no longer leaves any doubt that the increase of reactivity was in reality a nuclear explosion, only slower than one taking place in a nuclear bomb. In a nuclear bomb the cladding of uranium-235 is assembled in such a way that the 'critical mass' that is being created during the detonation causes a chain reaction and the release of heat in the course of some millionths of a second. In the reactor the chain reaction is supported by a special regulative régime. The result of speeding up reactivity 440-fold in one second is also an explosion, but a slower one taking place in a relatively big reactor. However, even the amount of uranium-235 in a reactor of the RBMK-1000 type, when the content is 2 per cent, equals almost 4 tons, which is enough for 270 nuclear bombs of the type dropped over Hiroshima. While the Hiroshima bomb contained 15 kg of uranium-235, there was no less than 2 tons of uranium-235 in reactor No. 4 in April 1986, although it had already been in operation for more than two years.

The British model of the explosion also fails to assign any role to hydrogen or carbon monoxide. When the temperature in the fuel elements is increased at the expense of the chain reaction on fast neutrons (prompt criticality) the fuel elements start to melt. This melting in itself causes a sharp increase of pressure in the reactor as the volume of melted materials increases by 10 per cent, reducing the volume occupied by gases (radioactive fission products) in the core. The spread of the melting fuel immediately destroys and deforms all the inner structures. At the moment of the explosion of the structures, according to the estimates, 57 per cent of the reactor's fuel was already melting and 10 per cent had melted. The collapse of the structure blows this burning mass from the fuel channels into the water, thus contributing to an instant steep increase of pressure and causing the entire reactor to explode. The model demonstrated that such an explosion rather than an instant expansion of steam took place in the course of less than a second (some tenths of a millisecond).

Most important in these conclusions was the fact that the heating up of the reactor took place at the expense of the accumulating flow of fast neutrons. In this case the heating-up process takes place so quickly that it is simply impossible to cool it down by the use of water. The water has not got sufficient thermal conductivity to absorb heat during such reactions. Reactions on fast neutrons take place at least 1,000 times faster than ordinary chemical and physical processes. The nuclear explosion, in the assessment referred to by the technical term 'prompt criticality excursion', was the first; the reactor explosion was an effect of it. The authors of the British assessment admit that 'if the reactor becomes critical on fast neutrons then a change in power takes place in periods of time so short that they exceed the speed of any control system. Consequently, the reactor becomes uncontrollable'.

Apart from these official assessments made by the American and British organisations in charge of nuclear energy programmes, a number of specialist seminars and discussions were held in both countries, in an attempt to provide an independent critical analysis of both Soviet and Western official estimates. Interesting seminars were organised by British and American Nuclear Energy Societies (the Nuclear Society of the USSR was established only in 1989). Materials from the British Nuclear Energy Society's seminar on Chernobyl were published in 1987.[6] The assessment made by the British society assumes that during the reactor's second rapid increase of reactivity, local temperatures at the places where the fuel melted could have reached 4,000 to 5,000°C. These temperatures are undoubtedly no longer 'chemical'. In this case even a steam explosion is impossible, as the steam ionises and dissociates into gases – oxygen and hydrogen. These are detonating gases, but they do not explode at such temperatures. The English radiobiologist Don Arnott has called the explosion 'a nuclear explosion of the reactor type', stressing its difference from the explosion of a nuclear bomb, during which the temperature in the epicentre reaches a million degrees. In Arnott's view, if the reactor's structures had been bolted together, or covered by a special containment, the explosion would have been much more powerful – destroying everything surrounding it.[7]

The international nuclear industry's reaction to Chernobyl

Both the American and the British accounts, which described every technical defect of the Soviet model RBMK-1000, every mistake made by the operators and the way they ignored the security measures, gave a detailed explanation as to why nothing similar could happen when operating reactors at nuclear power plants in the United States and Great Britain. However, the only unique Soviet construction flaw in the RBMK reactors was the wrong position of the graphite displacers of the control rods for emergency protection. All the remaining defects of the Chernobyl reactor were indeed impossible to find in combination in any Western or Japanese model. But taken separately, the slow shut-down of the reactor due to the excessively large size of the core, the use of graphite moderators of the neutrons and the positive void reactivity coefficient were factors present in some Western models as well.

The most common reactor model not using graphite moderators, of the type being used when the then-maximum accident took place at Three Mile Island in 1979, indeed proved immune to reactivity accidents of the kind that took place in Chernobyl – caused by runaway fission chain reactions of uranium-235. In such a PWR reactor, which corresponds to the Soviet water-moderated energy reactor (VVER-440 and VVER-1000), the water functions both as a moderator of the neutrons and as a coolant. Therefore, the loss of the coolant, for one reason or another, also represents a loss of the moderator of fast neutrons. The reactor is constructed in such a way that the fast neutrons alone cannot support a fission chain reaction. To get a sustainable chain reaction on fast neutrons a far more compact deployment of the nuclear fuel is required. This is possible only in special so-called 'fast' reactors, in which the coolant is liquid metal.

In reactors of the VVER-1000 type the size of the core is almost three times smaller than in RBMK-reactors, and therefore the reactor can be stopped by protection control rods in the course of 3–5 seconds (even in one second in reactors of this type in Japan). However, after Chernobyl all reactors fitted with graphite moderators and large cores could be categorised as 'dangerous' in the event of a maximum accident. An accident of the Chernobyl type that was earlier conceived as impossible took place due to a loss of the coolant (water or compressed gas) and the simultaneous failure of the accident protection rod system. As the graphite moderators remained, a rise of reactivity also took place. However, even without such an increase the sudden loss of the coolant could in this case have led to the meltdown of the core owing to the 'operating' flow of slow neutrons, much faster than the time it would take to switch on the emergency cooling systems. Previously, the melting of the fuel elements was considered possible only because of the residual radionuclides, and not as a result of neutron emission during the fission of uranium-235. Now it was also necessary to add the neutron danger to the radionuclides' alpha, beta and gamma radiation. The residual radionuclides heat up the fuel elements during a period of time measurable in minutes, and all the emergency cooling systems are adjusted for this. Even the very fastest among them could not have been switched on in less than 10 seconds. For neutrons even 10 seconds is slightly too much. This simultaneous loss of the coolant and failure of the control rod system was earlier considered highly unlikely in American and British reports. After the Chernobyl accident, however, experts were no longer able to confirm that this was completely impossible.

The end was therefore inevitable for all reactors fitted with graphite moderators. The nuclear industry did not admit it straight away in 1986 – that would have caused too many economic and political problems. But gradually this 'sentence' was carried out. As early as 1986 all nine military reactors at the Hanford reservation in the State of Washington on the west coast of the United States were stopped. Construction-wise, these reactors most closely resembled the Chernobyl type. The Americans held the view that the first Soviet military type reactors constructed by I. Kurchatov were copies of the American military reactors. Later this 'military' model was adjusted in the USSR to produce electricity. In military reactors it is not necessary to create high pressure and produce steam at all. It is also possible to make do with uranium without enriching it with isotope-235. Under these circumstances graphite is a far better moderator than water and facilitates a quicker turn-out of plutonium – which was the main goal. Subsequently a decision was made to stop operations at the Hanford site completely and dismantle and bury everything. (This takes thirty years and costs nearly a hundred billion dollars.) The Soviet equivalent to the American centre is situated in Kyshtym in the southern Urals, and the five reactors there have also been stopped.

Great Britain was worse off. The main British reactor model for nuclear power stations is fitted with graphite moderators and the size of the core is very large. (The core is 10 metres high, which makes it comparable to the RBMK reactor.) The cooling and the absorption of heat from the fuel elements is carried out with the help of compressed carbon dioxide, making it possible to reduce the pressure in the reactor (19 atmospheres in old models, 40 atmospheres in new ones), which in turn strengthens its safety. (In water-moderated reactors the pressure in the core reaches 160 atmospheres.) Old versions of such reactors –

'Magnox' – that were constructed in the late 1960s and early 1970s have also recently been stopped. It will take an estimated twenty-five years to demolish them, and this could prove more expensive than the construction of new reactors.

New reactor models fitted with graphite moderators and gas cooling systems which were built in the late 1970s and in the 1980s remain in operation, but the construction plan for new nuclear power plants is to be postponed for a review by 1995. In connection with the British Government decision of 1989 to freeze the construction plans for nuclear power plants, Lord Marshall – the main supporter of this programme – has resigned. During the privatisation of the electricity industry, which had been prepared by the Conservative government, it became obvious that the public would not buy shares if the 'package' was to include the twenty-nine British nuclear power plants, producing approximately 20 per cent of all electricity. For this reason, the government is compelled to keep all the nuclear power plants in the public sector.

Only in 1990 did France endorse a decision to close down all its reactors fitted with graphite moderators that had been built before 1970.

The Chernobyl accident also came as a heavy blow for all the Western plans to construct reactors based on fast neutrons. Interest in these reactors emerged because they allow for the immediate use of the plutonium being generated in them as nuclear fuel. But the switch-over to plutonium as fuel is a difficult problem, as plutonium is in itself a nuclear explosive. Further, the regeneration of plutonium from processed fuel requires advanced radiochemical processing and is accessible only to countries having nuclear arms. The American assessment of the Chernobyl accident admitted that American experimental reactors on fast neutrons also have a positive reactivity void coefficient. In fast reactors there are no neutron moderators. The heating up of the coolant is directly caused by fast neutrons in a controlled fission chain reaction of uranium-235 or plutonium. By 1986 a reactor on fast neutrons was already under construction for a long time at Clinch River, USA. It was to have had a power of 380,000 kw (and cost an estimated US$3.6 billion). But after 1986 construction was abandoned.

Britain also totally stopped its construction and research scheme for fast reactors in 1989, despite the fact that by 1988 £4 billion had already been spent on it. This was the end of the programme, adopted in 1975, which envisaged that one third of all nuclear electricity would be produced by fast reactors by the year 2000. In 1990 Britain also refused to take part in and to continue to fund the joint European fast-reactor project. A special Energy Commission set up by Parliament, consisting of members from the three main political parties, came to the conclusion that fast-neutron reactors could not become profitable before the year 2020. By that time current projects would already be outdated. At present only France and the Soviet Union are continuing work on projects for industrial fast reactors. The French fast reactor 'Superphoenix', with a power of 1,200,000 kw, is the biggest in the world among fast reactors. But at the moment it is considered a 'prototype', as energy produced by it is twice as expensive as energy produced at ordinary nuclear power plants. Besides, most of the time it is out of operation due to the high break-down rate of the cooling system on liquid sodium.

Of the 149 reactors in the world that were being planned or under construction in 1986, fifty-three had already been abandoned by 1989. Moreover, not only the nuclear plant construction industry but also the research in this field began to decline. The first country to reduce allocations for research on nuclear energy was the United States (in 1989–90). Only in Japan has a research budget for nuclear energy in 1991 been increased – 306 billion yen is allocated (an increase of 3.4 per cent compared to 1990). A considerable proportion of the nuclear budgets in other countries is now allocated to research not on new reactor-types but on safety systems and how to decommission, dismantle and bury outdated nuclear power plants. It is not unlikely that by 2010–2020 the number of nuclear power plants that have to be dismantled will exceed the number of nuclear power plants under construction. So far the major exporters of reactors for nuclear power plants have been the United States, the Soviet Union, France and Japan. If the current trend continues, Japan could gain a monopoly in this field as well in the coming century. The Soviet export reactors – VVER-440 and VVER-1000 – are, as far as their latest models are concerned, fully competitive compared to Western models when parts of their equipment are modernised. They require emergency stops more rarely and therefore their productivity is high.

However, the countries that have Soviet reactors are already 'saturated' with nuclear energy. At present 49 per cent of all electricity in Hungary is produced by nuclear power plants; in Bulgaria the percentage is 36, in Finland 37 and in Czechoslovakia 26. Japan produced 28 per cent of its electric energy at nuclear power plants and intended to increase this level to 50 per cent by the year 2000. Japan's neighbours – South Korea and Taiwan – were already close to reaching this level of nuclear electric energy (49 and 48.5 per cent respectively) and it was precisely the development programme for nuclear power stations that made possible the rapid industrialisation of these countries.

In Europe, apart from France, Belgium (67.3 per cent of its energy generation), Sweden (46.1 per cent), Switzerland (38 per cent), Germany (32 per cent) and Spain (31 per cent) all rely heavily on nuclear energy. In most of these countries, apart from Sweden, there were no serious alternatives to nuclear energy. Their transition to nuclear energy began as a result of the oil crisis in 1973–4. The new increase in world prices of oil in 1979 (up to US$40 a barrel) weakened the impact of the 1979 Three Mile Island accident in the United States on the European programmes. Nuclear energy simply replaced oil, as oil was too expensive.

In 1986, after the Chernobyl accident, a peculiar 'nuclear crisis' occurred. The introduction of new safety systems and higher insurance inevitably increased the cost of electricity generated by nuclear power. This was accompanied by a swift drop in oil prices, which in turn also reduced prices for gas and coal. Towards the summer of 1986 the price of oil had fallen to 10 dollars a barrel, which, when inflation is accounted for, represented a return to the pre-crisis level of 1972. This allowed some countries to drop nuclear energy, which had become unpopular, and return to traditional energy sources. Austria decided not to put into operation its first nuclear power plant, which had already been built. Italy stopped two nuclear plants and cancelled several projects. One plant was also closed in Spain, not far from Barcelona. A couple of plants were either halted or not put into operation in the United States. The apparent end of the Cold War promised a further reduction in demand for oil, as the military-industrial complex and the armed forces are the biggest users of liquid fuel and given priority in using it.

In 1990, however, a new oil crisis emerged and absolutely no prognoses are now, in 1991, capable of predicting the fate of nuclear energy in the coming decade. Given these circumstances, it is of particular importance that the technical aspects of the Chernobyl accident be made known as precisely as possible. So far everything that is related to the Chernobyl accident is in the form of preliminary assessments 'for official use'. Attempts made by more or less qualified amateurs to present a realistic picture of the accident and its ecological and medical implications are also only at a preliminary stage.

What does the sarcophagus reveal?

It is well known what the sarcophagus or 'ukrytie' (shelter) as it is officially termed, covers. It covers the ruined reactor block No. 4 which contains approximately 95 per cent of the remaining nuclear fuel, long-lived radionuclides and plutonium that was made in the reactor over 865 days – until 25 April 1986. According to all the estimates, some 4 per cent of the nuclear fuel was blown out of the reactor and up into the air during the accident. However, this figure takes into account only the radioactivity which spread in the form of a plume of gases and aerosol particles for ten days after the reactor had exploded. The real outburst was more substantial. Big pieces of the reactor fuel and of graphite, fragments of the fuel elements etc., which were thrown up onto the roof of the machine hall, at the roof of reactor No. 3 and over the square surrounding the accident reactor, were collected during the following days by soldiers of the chemical troops and thrown back into the reactor-crater. According to the eye-witness description given by the commander in charge of this operation, Major General Nikolai Tarakanov, published in 1990, more than 3,000 soldiers voluntarily took part in removing the pieces of radioactive fuel from the roof of energy block No. 3 and the machine hall. Equipped with spades and special grabbing devices, they threw pieces of the nuclear fuel elements and graphite back into the crater of reactor No. 4.[8]

Judging from this eye-witness report, the cleaning of reactor No. 3 continued for at least twenty days in September 1986, and a record amount of three tons of radioactive materials was removed in the course of a single day. This operation was undertaken before the sarcophagus was constructed. Even earlier, in May, pieces of radioactive fuel and graphite were collected from the square surrounding the reactor. According to the descriptions, the total quantity of this material came close to 100 tons, which is about 5 per cent of the mass of the core and the graphite. Therefore, physicists currently say that the sarcophagus covers no less than 90 per cent of the reactor's total radioactivity.

In 1990 the leading scientists S. T. Belyaev, A. A. Borovoi and A. Yu. Gagarinsky of the Kurchatov Institute of Nuclear Energy presented a paper at the IAEA on the condition of the nuclear fuel remaining in the reactor. This paper (IAEA–SM–316/90, which is so far available only in a typewritten copy in English) was the result of investigations undertaken by a special permanent complex expedition set up to study everything taking place inside the sarcophagus. As a result of these investigations it has been revealed that the reactor pit is practically empty. All the graphite has burnt up and the reactor fuel, having melted down, has spread in the form of lava and is distributed in the reactor's lower premises. Some 5,000 tons of sand, clay, dolomite, lead and other materials spread by helicopters between 27 April and 2 May 1986 are admitted to be virtually useless, as the upper 2,000-ton plate raised over the reactor crater prevented the protective materials from directly hitting the crater. This crater has so far not been covered. The paper thus refutes the still-repeated assertion that precisely 'as a result of the covering of the accident reactor, the reactor pit was covered by dry masses, and already on 6 May the emission of radioactive substances from it practically came to an end'.[9] In reality, the graphite fire stopped only when virtually all the remaining graphite had burnt up.

In the middle of 1990 one of the authors of the paper given at the IAEA, A. A. Borovoi, published a small booklet called 'Inside and outside the sarcophagus'.[10] This booklet, consisting of sixteen pages of text and a few interesting illustrations, was printed in only 250 copies. Apparently for this reason it remains unknown and uncommented upon in the general press. For the first time it gave fairly detailed information on the condition of the remaining nuclear fuel and on the damaged reactor itself. At present some 750 kg of four different isotopes of plutonium remain in the reactor. Of these, 420 kg are plutonium-239, with a half-decay of 24,000 years. As far as radiotoxicity is concerned, plutonium emitting alpha particles is the most dangerous. Apart from plutonium, 81 kg of caesium-137 with a total radioactivity of 6 million curies remain. So do 43 kg of strontium-90 with a total radioactivity of 6 million curies. A. A. Borovoi admits that during the helicopter drop of sand, dolomite and other materials 'if the dropped materials hit the reactor, then only a small part of them hit the reactor pit'.[11]

The most interesting discovery of the investigations undertaken inside the sarcophagus in 1988–89 was that the reactor explosion not only pushed the upper plate and fuel into the air, but also destroyed the lower plate.

The reactor space proved to be virtually empty – any significant fragments of cladding in it are missing. The upper roof of the reactor, weighing approximately 2,000 tons, stands on a rib at an angle of 15 degrees vertically, on the one side leaning towards the edge of a metal tank, on the other on a reinforced concrete plate resting on this tank. A considerable number of cut-off technical pipes run from it. The lower roof of the reactor was pressed 4 m down after having smashed the massive metal constructions located in the premises under the reactor. The reactor's south-west quadrant is gone – it was destroyed during the active stage of the accident.[12]

This new information about the state of the reactor makes it possible to assess the character of the initial explosion differently. It also resolves any doubts concerning the possible quantity of radioactive materials blown out of the reactor. According to the previous scenario, the explosion that took place in the lower part of the reactor was so powerful that it lifted the upper plate, with a weight of more than 1,000 tons, into a vertical position. But according to all the simplest assessments such an explosion should have blown out far more than 4–5 or even 10 per cent of the reactor fuel, especially as the fuel assemblies were attached to the upper plate of the reactor construction. Judging from the real description of the reactor, which is only now emerging, the lower plate, i.e. the bottom of the reactor, was destroyed a moment earlier, and therefore a large part of the inside of the reactor pit was pushed downwards, not upwards.

The Soviet State Committee on the Use of Nuclear Energy presented detailed technical descriptions of reactors of the RBMK-1000 type (Appendix 2. The construction of the reactor installation) at the IAEA meeting held in Vienna in 1986.[13] According to this description, the reactor's upper plate has the shape of a cylinder 17 metres in diameter and with a height of 3 metres. The upper and lower parts of this cylinder are made of 40 millimetres thick solid steel. Inside the cylinder there were vertical reinforced ribs. Numerous pipes for the technological and steering channels crossed each other. The plate – or rather the metal structure – was not fastened to the vessel of the reactor pit, but rested on sixteen rollers. It is precisely this plate that receives the pressure of the weight of the functioning channels loaded with the fuel elements, the deck of the central hall and the pipelines of the water and steam communications. The lower plate with a diameter of 14.5 metres and a height of 2 metres was loaded with graphite cladding with supporting nodes and the lower water channels. The supporting metal construction, on which the lower metal construction is fitted, consists of steel with enforced ribs 5.3 metres high, crossing each other perpendicularly in the centre of the reactor.

The cylinder-like reactor container itself with an outside diameter of 14.52 metres and a height of 9.75 metres, made of 16 millimetres thick steel, forms an enclosed reactor space from the upper to the lower plate. But this cylinder container is not welded or fastened together firmly from the upper and lower metal constructions. Another cylinder (a tank for lateral biological protection) with an outer diameter of 19 metres and an inner diameter of 16.6 metres made of 30-millimetre-thick steel surrounds the reactor. The inside of this construction is divided into sixteen vertical hermetic modules, filled with water, the heat of which is removed by the cooling system. It is on this biological protection tank that the upper metal construction, standing vertically, presently rests.

Judging from this description, a substantial part of the reactor fuel was pressed down into the various rooms and corridors under the reactor. In as much as the upper and lower plates differ in thickness and weight it is unlikely that they would have given in simultaneously, when the pressure in the reactor sharply increased as the explosion developed in the course of the 4.9 seconds after the scram button had been pressed. The upper plate, apart from its own enormous weight, also had a heavy deck for biological protection. More than 1,000 pipelines and metal constructions running through the plate resisted its upward movement. The pressure must have been enormous to tear these steel pipes apart.

Enormous power was also required to crush all the metal bearings on which the lower plate was placed. If both the plates had been fastened to the reactor vessel, then this vessel, made of thinner steel, would have been destroyed. But it also strengthened the biological protection cylinder which fulfilled some of the functions of containment. As a substantial part of the reactor fuel was pushed down into the premises under the reactor where it now fills a large area, it is perfectly clear that the lower metal construction plate was first pressed four metres down. However, while being pressed down, it should have functioned as a valve reducing the pressure inside the reactor. Under these circumstances the upper plate could be pushed up only in the event of a new increase in pressure. It may therefore be assumed, as a hypothesis, that the lower metal construction was pushed down (four metres, as shown by A. A. Borovoi) when the power in the reactor first increased 100-fold, having destroyed the reactor core and the fuel elements attached to the upper roof.

The graphite blocks were mounted precisely onto the lower metal construction. Therefore they crashed down together with a considerable part of the reactor fuel. In various parts of this mixture constrained by the explosion, new centres of 'criticality' and a chain reaction on fast or slow neutrons would have occurred. Thus, there was a danger that the mass not only of uranium-235 but also of plutonium would reach a critical level. The volume of the critical mass for plutonium is several times smaller than for uranium-235. This caused a new 440-fold increase of reactivity, as measurements of the explosion show, and the lifting up of the upper plate and the blow-out of fuel. The explosion dispersing the nuclear fuel also simultaneously stopped the chain nuclear reaction.

Up to now the most detailed description of the explosion of the Chernobyl reactor has been given in Grigory Medvedev's documentary novel *Chernobyl Notes*.[14] The author is an engineer by profession, specialising in the construction and maintenance of nuclear power plants. He has interviewed several operators, 'liquidators' and experts, and was in Chernobyl after the accident. Although his novel is not, of course, a purely scientific piece of work, the author tries as far as possible to explain precisely the causes and character of the reactor explosion and all the events that took place during the first days after the accident. Knowing of the Chernobyl Scala computer's last print-out before the accident developed, Medvedev assumed that the explosion took place in the upper third of the core. As described by him, fifty tonnes of fuel were pushed upwards from the reactor (in the form of gases and aerosols). Seventy tonnes were spread in various directions in the form of fairly large pieces and only fifty tonnes remained in the reactor pit. The blow-out took place as a result of the combustible gas (a mixture of hydrogen and oxygen) accumulating as a result of the overheating of the nuclear fuel and the dissociation of the water. At present, as a result of the complete revision of the computer models, nuclear physicists, as we have seen, have come to the conclusion that the initial explosion (the 100-fold increase of reactivity and the subsequent 440-fold increase) took place in the lower part of the reactor. But while this happened, not seventy tonnes of the fuel was blown out, as suggested by Medvedev, but twenty tonnes of the uranium fuel at the maximum. This can be explained only by the fact that the sharp increase in the pressure inside the reactor first destroyed the bottom of the reactor and only a second later pushed away the upper plate. The second explosion obviously further damaged some of the lower premises, having created a draught of air through the reactor pit, which in turn supported the graphite fire during the next few days.

The 440-fold increase in reactivity as a result of the local heating up of reactor fuel to 3,000–5,000°C at the expense of fast and slow neutrons and the runaway chain reaction that followed was without doubt a nuclear explosion. The steam contained in the reactor is as a result transformed into hydrogen and oxygen as the dissociation of water starts at a temperature of 2,500°C. However, this high temperature prevents the combustible gas from exploding, as water is being formed from the mixed gasses. This can happen only when the combustible gas leaks out and is subject to cooling. Gas explosions no doubt took place during the accident, but only after the upper plate had been pushed up and only after the approximately 1,900 steel pipes running through it had been destroyed. In itself, combustible gas is incapable of producing explosions sufficiently powerful to squeeze and flatten the lower metal constructions and blow up the upper ones, having lifted up a 2,000 ton plate. The combustible gas creates sound effects and may cause a fire, but its explosive power is too small and in the course of 3–4 seconds many products of chemical reactions simply could not be developed and accumulated.

As a photograph of the reactor's upper plate taken by periscope (Borovoi's drawing No. 6) demonstrates, at present the remains of the pipes and other communications that were damaged during the explosion are hanging from this plate. Judging by its present position (an incline of 15 degrees in the direction of reactor block No. 3), a large part of the reactor fuel and graphite should have been ejected in this direction. This obviously also explains why so many pieces of the fuel elements had to be removed from the roof of Block No. 3.

After the sarcophagus had been built, a number of statements were made saying that any new warming up of the fuel in the remaining parts of the reactor was absolutely impossible. The appearance somewhere inside the sarcophagus of centres of 'critical mass' and the release of neutrons were completely excluded. But these statements were made for the public. In reality physicists were not convinced and did not exclude the possibility that due to some new dislocations or crumbling, new fires and melted fuel could cause the melt-down of concrete by the remaining reactor fuel (i.e. the 'China syndrome' could take place). When, in 1988, physicists for the first time were able to undertake work inside the reactor, their first task was to prove the 'subcriticality' of the fuel. Borovoi reports that only 'in the spring of 1989 was the State Committee informed of the absence of nuclear danger of the site'.

However, a new danger had emerged: the gradual brittlement of the lava, forming masses of fuel, mixing and floating together with sand, concrete and metal due to the high temperatures. The decay in material subject to powerful radiation has long been known, and it is precisely this which limits the operative period of the reactors. The lava covering the inside of the sarcophagus gradually turns into dust and a corrosion of the metal pieces also takes place. Any collapse threatens the spreading of dust into the air and the outburst of radioactivity from the sarcophagus. The sarcophagus is by no means hermetic and has approximately 1,000m^2 of various openings and cracks.

The future

Physicists so far guarantee 'that the condition of the structures inside [the sarcophagus] is safe for another 10 to 15 years'.[15] Thereafter it will be necessary to demolish the entire complex by safe methods and undertake a reliable reburial of the remnants of the reactor. This path was followed by the Americans with their reactor that was damaged in 1979 at Three Mile Island. Such work is probably being planned, as at present there are no robots which could carry out the necessary operations. In the case of the USA the role of the sarcophagus is successfully being filled by a protective hermetic containment-cap that was installed at the nuclear power plant prior to the accident. Periscopes have been inserted through it to monitor the reactor, and only recently people were able to enter. It has been found that the melting of fuel was far more extensive than assumed initially. Approximately half the entire core has melted and about a third of the total melted material has leaked down to the reactor floor. More than 100,000m² of contaminated water which also needs to be removed is kept in the premises under the reactor.

In the Soviet Union a discussion is now going on as to whether the same should be done at Chernobyl or whether some kind of super-sarcophagus, provisionally named 'ukrytie-2', ought to be built above the existing one, solid and hermetic enough to withstand any collapse of the construction inside sarcophagus No. 1. 'Ukrytie-2' will be needed for several centuries, and then our descendants will decide what should be done to the plutonium remaining inside. Radioactive strontium and caesium, whose half-life is approximately thirty years, will disappear after 180–200 years, and this will ease the work of our descendants and their robots at Chernobyl.

Any accident, apart from its technical, medical, ecological and other aspects, also has an economic side to it. What are the economic costs of the Chernobyl accident? In the United States it has been estimated that the final bill from the accident at Three Mile Island will cost the consumers approximately US$130 billion. As the Three Mile Island plant is not state property, the bill will be covered through money collected from consumers. As a comparison it should be said that in the 1970s the construction of one reactor in the United States cost US$0.5 billion. Lately, due to new safety requirements and delayed construction, US$3–4 billion have been spent on the completion of already started but not yet completed reactors. In the USSR, Yu. I. Koryakin recently estimated that when not only the technical, construction and various indirect expenses are taken into account, but also the medical and agricultural programmes 'for the liquidation of the consequences of Chernobyl', the disaster will have cost the USSR approximately 170–215 billion roubles by the year 2000.[17]

Prior to 1986 the construction of one reactor in the USSR cost approximately 0.5 billion roubles. After 1986 the cost rose immediately. In the USSR the above-mentioned 200 billion roubles are borne by the entire economy and not just by electricity consumers in one state, as is the case in the USA. However, the expenditure will evidently continue into the coming century. And even in future centuries those of our descendants who have the task of dismantling supersarcophagus No. 2 or building a supersarcophagus No. 3 will remember that the people who invented such an original emergency protection system, which starts and increases reactivity for a few seconds before it can shut down the reactor, ensured that these descendants had more than enough to keep them occupied.

Notes

1. See *Chernobyl'. Sobytiya i uroki* (Moscow: Politizdat, 1989), p. 38.
2. Report of the U.S. Department of Energy's Team Analyses of the Chernobyl-4 Atomic Energy Station Accident Sequence. November 1986. US Department of Energy, Assistant Secretary for Nuclear Energy, Washington, DC. 20585 DOE/NE-0076. (The paper consists of 6 parts and contains 8 appendices.)
3. *Novyi mir*, no. 6, 1989, pp. 3–108.
4. E. Adamov, V. Vasilevsky *et al.*, 'An analysis of the first phase of the development of the accident in Block Four of the Chernobyl Nuclear Power Plant', in *Atomnaya energiya*, 64, no. 1 (1988), pp. 24–7.
5. J. H. Gittus *et al.*, *The Chernobyl Accident and its Consequences* (London: United Kingdom Atomic Energy Authority, 1988), 2nd edition.
6. British Nuclear Energy Society (BNES), Seminar Report, *Chernobyl: Technical Appraisal* (London, 1987).
7. Ibid.
8. See the essay 'Chernobyl. Soldaty i generaly' in *Literaturnaya Rossiya*, 28 September 1990.
9. *Chernobyl'. Sobytiya i uroki*, p. 209.
10. A. A. Borovoi, *Vnutri i vne sarkofaga* (Moscow, 1990).
11. Ibid., p. 6.
12. Ibid., p. 11.
13. See *Avariya na Chernobyl'skoy AES i yeyo posledstviya*. Prepared for the IAEA experts' meeting, August 1986. Appendix 2, pp. 14–15.
14. *Novyi mir*, no. 6, 1989.
15. Borovoi, *Vnutri i vne sarkofaga*, p. 14.
16. Ibid., p. 15.
17. Yu. I. Koryakin, 'Tsena Chernobylya – 200 milliardov', *Energiya*, no. 8 (1990), pp. 2–6.

12 Glasnost, perestroika and eco-sovietology

Igor I. Altshuler, Yuri N. Golubchikov and Ruben A. Mnatsakanyan

Today the USSR is in a state of severe environmental crisis. The basic reason is that for more than seventy-three years the Marxist-Leninist methodology of using violence to redistribute wealth has been applied both in social and economic spheres as well as in environment and resource management. This methodology posits a strong central power with the monopoly to make decisions on such matters as shifting nationalities from place to place or reversing the natural course of rivers. And since the USSR accounts for one sixth of the planet's land surface, the process of decision-making (especially strategic decisions) and management in the USSR has become a global environmental factor. Thus, the state of the environment in the USSR and consequently the world will largely depend on the Soviet Union's future mechanisms of decision-making (especially strategic ones) and management.

So the destruction of nature and the environment in that part of the biosphere which belongs to the USSR is not only an internal affair. If this devastated and under-nourished country with its powerful military-industrial complex were turned into an ecological disaster zone, it would be indeed dangerous for the whole of humankind.

Another reason why the state of the environment in the USSR is of global importance is that its territory covers the high latitudinal peripheries of the biosphere. These affect the other parts of the biosphere more than vice versa. In fact, the USSR, like a continent, has many northern-type fragile ecosystems with very little capacity for recovery. And they are overfilled with the most dangerous technologies of all – chemical, biological and nuclear – which, furthermore, are not only poorly designed and serviced but inadequately controlled.

Seventy-three years of misinformation (or incomplete information) on the environmental situation in the USSR seem absolutely unacceptable, given that the USSR disposes a major part of the world's natural resources. The global environmental effects of the Soviet centralised system, especially the decision-making processes, seem to be underestimated in the West. 'Environmental Sovietology' ('eco-sovietology') is rather underdeveloped, compared, for instance, to 'military sovietology'. But today, with less military confrontation, it is necessary to pay more attention to the Soviet system as a global environmental factor no less significant and dangerous for the rest of the world than the military factor.

Contribution to eco-sovietology

One of the most significant contributions to 'eco-Sovietology' is *The Destruction of Nature in the Soviet Union* by Boris Komarov (pseudonym of Zeev Wolfson).[1] It was written in the USSR clandestinely and sent illegally to the West where it was published in 1978. This happened just at a period when practically all the experts of the USSR Academy of Sciences and universities in economics, geography, ecology or resource management were occupied in 'research' aimed at approving various projects to 're-make nature': reversing river courses, constructing unnecessary canals or railways, or implementing large-scale devastating amelioration and chemicalisation programmes, etc. In his book, Komarov emphasised the necessity of interdisciplinary research (especially oriented to solving environmental problems) and the restructure of the system of science organisation in general so that it was no longer aimed mainly at long-term military/political projects. Komarov stressed the idea of the inadmissibility of violence towards nature and society as a means of solving various problems and implementing economic and political projects. He also stressed that the USSR lacked adequate laws and that both the elaboration and implementation of laws was in the same hands.

Komarov's book is a good example of a holistic approach to the USSR's environmental problems. It is by no means limited to 'nature conservation' aspects, considering as it does the environmental problems in the USSR in a broad context of political and economic affairs including the role of the Communist party leadership and the military-industrial complex in the 'destruction of nature'. One of the author's important conclusions is that the principles of the Communist party-directed planned and 'nature-intensive' economy are inimical to the interests of the environment and human life.

Komarov was one of the first authors to attract attention to the planetary effect of the destruction of nature in the USSR (for instance, clearing the Siberian taiga and the danger from nuclear power stations). And this was written long before Chernobyl or glasnost.

This seminal book anticipated many of the fundamental ideas of 'new thinking' and perestroika. One can only conclude with regret today how harmful was the non-publication of Komarov's book and how many years have been wasted by the Soviet experts in environmental science and resource management. Boris Komarov's book touched on many problems but many of them have still not been analysed in detail. In addition, since the book was written long before the changes in the Soviet Union, it has no analysis of the perestroika period and thus, despite what has been published since, serious and systematic eco-sovietological research has still to be undertaken.

Environmental crisis in the USSR

The environmental crisis in the USSR is not just due to errors in policies and planning but to the very nature of a centrally planned economy which, we are deeply convinced, is environmentally destructive and dangerous. Major ecological disasters have been caused by the elaboration and implementation of strategic economic programmes. The drying up of the Aral Sea is, for instance, an obvious result of the party line on irrigated cotton as the monoculture of Soviet Central Asia. Among other examples we could refer to wind erosion of soils and dust storms on a huge scale (due to the policy of ploughing Kazakhstan's virgin land – up to 30 million hectares in the 1950s), the destruction of river basin landscapes of the European Russian plain by flooding them for hydroelectric projects (approved by CPSU plenums and congresses) and, finally, the Chernobyl catastrophe (with the general over-concentration in the European USSR of nuclear power stations and other dangerous installations).

During the whole of Soviet history, certain strategies (or Communist party lines) have been implemented. And in all these seventy-three years, economic development has been carried out primarily:

- at the expense of the extensive consumption of natural resources (which on this vast territory seemed to be inexhaustible) and with a lack of price value for natural resources (according to Karl Marx's 'labour theory of value', anything which has not been created by human labour but has been 'presented by God' – air, water, land, etc. – cannot be evaluated in monetary terms);
- with a lack of adequate laws (including environmental laws);
- with permanent changing of the 'rules of the game' in the economy and the subordination of economic and governmental policies to Communist party 'slogans of the current moment', admitting the use of any means for achieving goals;

with a lack of democratic principles of decision-making in all spheres of life;

and, finally, with lack of reliable information on the state of affairs in the country.

According to Professor Boris Vinogradov's viewpoint (referred to in the London *Times* of 9 February 1991) 'by the middle of last year 12 per cent of the Soviet Union was an ecological disaster area. Now it is 16 per cent'. We think that although this 16 per cent makes up a large area of 1,382,000 square miles, in reality the ecological disaster area in the USSR is significantly larger. And this area is not limited to nine 'zones of ecological disaster in the USSR' which have recently been determined by the USSR's Supreme Soviet. Practically the whole USSR's territory can be classified as an ecological disaster zone, with very few exceptions. Not only Lake Baikal, not only Chernobyl, not only the Aral Sea or Kalmykia, not only the Volga basin, but many other areas as well can be regarded as ecological disaster areas. According to Alexei Yablokov, a corresponding member of the USSR Academy of Sciences and deputy chairman of the USSR Supreme Soviet Committee on Ecology, 'the crisis in the Gulf is caused by oil, but the amount of oil spilt in the Tyumen oil-producing region in the last few years is greater than the Gulf. There is a danger the world is going to forget the crisis here'.[2] Professor M. Lemeshev, a very prominent and competent Soviet expert in ecology, asserts that the list of the USSR environmental disaster zones where both nature and human health and survival are endangered should be expanded to include vast territories of Central Russia, the Middle Ob' lowland, Southern Urals, Chukotka, the Yamal peninsula, the Southern Ukraine, North Caucasian health resorts, Crimea, Riga Bay coastal waters and dozens of large and middle-sized cities.[3] In an interview with the newspaper *Sovetskaya Kul'tura*,[4] the USSR Goskompriroda chairman Nikolai Vorontsov indicated that vast polluted areas exist on the left bank of the Ukraine's Dnieper River including the Donbass, Krivoi Rog, Dnepropetrovsk area, Zaporozhye and Mariupol. These polluted areas combined are comparable in scale to the areas affected by Chernobyl.

Unfortunately, the official documents published in the last two years which are the first 'open' (non-secret) sources of data on the USSR's state of the environment fail to mention the above areas in their compilation of critically polluted regions. Neither Goskompriroda's report,[5] nor the Goskomstat statistical abstract[6] contain reliable information on the degree of environmental degradation in the USSR. Thus, more adequate information on the state of the USSR's environment is available from indirect sources such as the mass media, speeches of deputies of soviets and public environmental movements.

Primary information for eco-sovietologists

On 16 April 1990 Nikolai Vorontsov declared on the Soviet Central television programme 'Face the nation', 'We must switch from a sharp downfall to a slow deterioration of the situation'.[7] This is a grave assessment of the situation but, to effect even this minimal change, Vorontsov (the first Soviet non-party minister) must have a complete picture of the country's ecological catastrophes and it is questionable whether this is so. The present level of glasnost is still clearly insufficient. As Vorontsov has emphasised, the Soviet Parliament does not even have open data on the financing of the three most important branches of national security – defence and its industries, state security and internal affairs. Even he, the environmental minister, is unable to state precisely how much the USSR spends annually on ecology – 'nine to ten billion roubles' (an accuracy of plus or minus a billion!). Even he does not have exact data on 'annual ecological losses' – whether 43 or 90 *billion* roubles.

It is thus necessary to obtain a qualitatively new level of complete and reliable information; unfortunately, up to now all categories of information consumers (state bodies, citizens, researchers, etc.) lack it.

The seventy-three year history of the Soviet system is a history of systematic misinformation on the environmental situation in the USSR. It should be noted that for the most part environmental data have always been secret in the USSR. What environmental information was available was marked 'For official use only' and distributed in negligible numbers of copies (for example, *Reviews of the State of Air* (or *Water*) *Pollution in the Cities and Industrial Centres of the USSR* (in an edition of only 100–200 copies)).

The state of affairs has also been worsened by the fact that much other official statistical data, for instance, on the state budget and, particularly the balance of payments deficit, military expenditure and human morbidity, were published in an incomplete and misleading form, serving the aims of official propaganda. It is significant that large-scale maps in the USSR still belong to the category of 'closed' (confidential) material together with almost all sorts of aerial and space photo images. The inscription 'For official use only' successfully conceals information on the USSR's schemes of forest taxation and land-use plans as well as large-scale geobotanical, geological, landscape and nature-protection maps.

The problem of primary data low reliability and incompleteness seems to be one of the most important problems faced but, unfortunately, almost never discussed by sovietologists and by environmental sovietologists in particular. This problem does not refer only to quantitative data but also to substitution (or misuse) of concepts which is very common in Soviet everyday life and misleads both experts and ordinary citizens. What has always been called either 'meat', 'sausages', 'money', 'democracy', 'social security', 'the fight for peace', 'people's friendship', or 'environmental protection' have in reality never corresponded to their essence. The official Soviet understanding of all these notions (and, in fact, many others) has always been different from the Western understanding.

The misleading substitution (or misuse) of concepts clearly manifests itself in the field of environmental protection. 'Pollution control', for instance, is interpreted by Soviet officials only as 'measuring' pollution, whereas in the West it means both measuring and taking measures against pollution. So, if a Westerner, for instance, reads in *Moskovskaya Pravda* that 'the state of Moscow air is under control', he would understand that air pollution in Moscow is being coped with satisfactorily. He would be mistaken, however: it would mean that Moscow air quality would be only *measured* and no more.

Goskompriroda and Goskomstat reports

If we turn to the two official documents on the state of the environment in the USSR, to what extent are their data reliable? Do the documents under consideration reflect the whole breadth and depth of the political, economic, social and environmental crisis in the USSR? Is there any possibility of using these materials in a practical way and, particularly, in forecasting the state of the natural environment in the USSR and in the world at large?

The Goskompriroda report contains the list of cities with the highest levels of atmospheric pollution (citing the specific substances causing the high level of pollution). The report also includes tables with the mean monthly mortality rate of the adult population (unfortunately, for only three cities) and numerous maps: maps of cities with the highest level of atmospheric pollution and increased population morbidity; a series of Chernobyl maps (including maps of densities of caesium and gamma-ray pollution of the official contaminated zone and, most important of all, of other Ukrainian, Byelorussian and RSFSR areas affected by Chernobyl's pollution). The report also includes schematic maps of sulphate and nitrate deposition as well as maps of forest resources, etc.

For the first time it publishes brief information on public environmental movements and groups and some details of their activities.

Over 120 people were involved in preparing Goskompriroda's report. Thirty-five of them are members of the inter-ministerial editorial board with First Deputy Chairman of the USSR Goskompriroda V. Sokolovsky at its head. So many authors – some twelve-page chapters have up to twelve authors – to a very large extent makes this report anonymous and removes personal responsibility for the data's reliability and conclusions from the authors. It was, incidentally, V. Sokolovsky who was responsible for the great level of secrecy and intentional distortion of environmental information (particularly as regards Chernobyl) when First Deputy Chairman of Goskomgidromet (the State Committee on Hydrometeorology and Environmental Control of the USSR). Unfortunately, the Goskompriroda report seems to have been compiled in a hurry and not edited carefully with its sections and even subsections very often inadequately coordinated and with many repetitions and discrepancies.

One senior drawback is that its maps and tables are based on separate republics. Thus, in all the parameters given, the Russian Federation, accounting for one-seventh of the Earth's land surface, is compared to Estonia with an area 377 times smaller. The total amounts of pollutants are given for the USSR in general or are classified by ministries responsible for branches of industries, but not according to territorial and *oblast* sub-divisions. Unfortunately, many parameters referring to the pollution of water, air and soils are not comparable, which makes it impossible to have a general, holistic impression of the pollution of regions as a whole.

Neither the Goskompriroda report nor the Goskomstat statistical abstract have clear definitions of such terms as 'pollution', 'zone of ecological disaster', 'sanitary norms', 'catastrophic state'. It is not clear how 'polluted sewage' differs from 'non-polluted sewage' or 'ecologically dangerous' from 'non-dangerous' (or even 'safe').

The report could probably be criticised in the same way as other official government material about the state of affairs in the USSR in its various aspects. The usual obvious discrepancies between the data presented and the real situation manifest themselves. For instance, it is known that in 1980–85, according to Soviet statistics, there was an increase in nature-destroying (nature-consuming) industries. In the same period the annual production of electric power increased by 251 billion kwt/hours, gas by 208 billion cubic metres, chemical fertilisers by 8.4 million tonnes, synthetic resins and plastics by 7 million tonnes. Moreover, the USSR is using much more raw material per unit of national income than the USA: 2.3 times more oil and natural gas; 3.1 times more steel; 2.8 times more chemical fertilisers; twice the electric power.[8]

Thus the Soviet economy's consumption of natural resources is, according to the Goskompriroda report, excessively high and growing. Yet the report states that the total air pollution in 140 industrial cities had, however, almost halved since 1972. What could be the reasons for such a mysterious atmospheric improvement? Or is it pure fiction?

In the Goskomstat statistical abstract some materials and tables presented seem totally non-informative, if not absurd. For instance, according to a table on the amount of recycled waste and percentage of this category of waste (by types), 100 per cent of slag pig iron and pig iron is recycled.[9] This seems utterly improbable to Soviet town-dwellers who can confirm their doubts only by looking through their own windows onto the streets outside.

Again, according to one of the tables of Goskomstat's abstract, the city of Kemerovo, known for its extremely high level of pollution, gives off half as many emissions as Novosibirsk or the same amount as Mogilev, which seems absolutely improbable. Still more improbable seem the data from the table 'Average concentrations of pollutants in the atmosphere of individual cities', according to which Kemerovo has indices equal to Vilnius or better than Moscow, Mogilev, Leningrad, Alma-Ata, Astrakhan and many other cities. And how is an 'average annual concentration' determined if Kemerovo's basic source of pollution is sporadic emissions (on these occasions the population says, 'They've released a huge amount of dreadful gas again' and the air in the city becomes bright orange) – we, the authors of this paper, witnessed such episodes creating a totally abnormal situation *for several hours*? So the annual *average* concentration of pollutants may be misleadingly lower.

It is not only public health which suffers from air pollution in the USSR but the actual life span of the population. According to Vorontsov's interview in *Sovetskaya Kul'tura*, the Soviet Union is fiftieth in life expectancy out of fifty-two countries which have such statistics. A disturbing situation. Furthermore, the average life expectancy of the USSR's population has been declining over the past twenty–twenty-five years. Particularly disturbing data on the health situation comes from the Aral area and the industrial regions of the Urals, southern Siberia, the Ukraine and the Far North.

The authors of the two documents refer to a list of the sixty-eight worst-polluted cities in the USSR and observe that some forty million people live in them: one out of every five urban dwellers in the USSR. It should be noted, however, that according to the Goskomstat abstract, observations on atmospheric pollution levels are conducted regularly in 421 cities with a total population of over 113 million people. Since there are more than 2,000 cities and towns in the USSR, the atmospheric pollution is obviously not measured ('controlled') in all of them. Unfortunately, the official documents considered do not cover even the available information. The tables cited above, for example, do not provide data on such severely polluted cities as Kharkov, Vladivostok, L'vov, Chernovtsy, Semipalatinsk, Chita and many others.

It is interesting to compare the list of the sixty-eight most air-polluted cities with the morbidity table on the adult population by individual cities. It turns out that cities such as Berezniki, Voskresensk, Gomel, Dzerzhinsk, Kremenchug, Leningrad, Tiraspol, Mogilev, Murmansk, Rovno, Simferopol, Kherson, Cherkassy, and Yalta, all of which have particularly high rates of oncological diseases and illnesses of the upper respiratory and digestive systems, are not even on the list of the sixty-eight cities with high air pollution rates. Also omitted is Norilsk, which annually emits one million tonnes of sulphur among its total 2.3 million tonnes of pollutants.

One of the favourite methods of Soviet statistics to mislead data consumers or to hide the truth on the real state of the environment is the use of summary data for estimating environmental pollution. The Goskompriroda report's table 'The emission of harmful substances from stationary sources in the atmosphere by individual cities' is a good example of such an approach. Yu. Izrael, the USSR Goskomgidromet Chairman, for example, reported in a lecture in 1988 at Moscow State University (Faculty of Geography) that the *total amount* of emissions of air pollutants in the USA was higher than in the USSR, testifying to the greater effectiveness of environmental protection activity in the Soviet Union. Is it possible, however, to measure the amount of emissions of sulphur and dust, nitrogen and radionuclides and publish summary data? Does it make any sense?

One significant factor is that the USSR's ecological services are by and large not admitted to monitor territories occupied by the military. The reports under discussion therefore exclude information about the area appropriated by the army for military bases, including testing ranges and 'closed zones', nuclear test sites of Semipalatinsk and Novaya Zemlya, etc. The reports also omit the subjects of poaching in places inaccessible to inspectors and the sale of arms to the local population for poaching as well as for use in inter-ethnic conflicts. People in the USSR can live next to an ecologically dangerous firing range yet consider they are living in an ecologically pure zone.

The ecological context of perestroika

Let us ask two questions. What does the world mean by perestroika? And why have Mikhail Gorbachev's policies and reforms been supported practically everywhere and by everybody both materially and morally – and even been rewarded with the Nobel Peace Prize?

The answers seem to be obvious. For the rest of the world perestroika means more democracy and freedom for the USSR and the Eastern bloc, withdrawal from Afghanistan, the destruction of the Berlin Wall and Iron Curtain, the lessening of military confrontation, and so on.

One of the core points of perestroika is glasnost, i.e. greater openness of some sides of Soviet life which used to be hidden before. Glasnost in the USSR has brought dramatic changes in Soviet citizens' minds. Having learned the truth about the realities of life, the Soviet people are reaching the conclusion that they can no longer tolerate their pattern of life because it is their *survival* as human beings which is on the agenda. The whole country is now in a state of ferment and it emerges that the majority of political events and 'social explosions' in the USSR have very serious underlying environmental origins or at least contexts. We are deeply convinced that the more glasnost in the USSR, the more the inhabitants become concerned about their survival. It is this which produces the main difference between Eastern and Western environmentalists: in the West they are fighting mainly to *protect nature*; in the East they are fighting to *protect themselves*, to survive and to eliminate the political and economic reasons of the environmental crisis.

It has to be noted that to a very large extent the current 'wars of laws' or 'wars of sovereignties' between central and regional governments in the USSR have environmental (resource) undercurrents. Many leaders of influential popular fronts in the Baltic states and Armenia (the 'Karabakh' movement) started their political activities as environmental leaders, while the popular fronts themselves grew out of ecological demonstrations and rallies. They are fighting against destructive all-union projects, as well as for the chance to eat sausages other than of Chernobyl meat (confirmed in several reports), toilet paper or rats (as claimed in the mass media), or at the very least to know what their food is made of. They would prefer not to live on radioactive waste burial sites or have to find that they do so – as was the recent case in Moscow where three secret nuclear waste burials of the 1940s and 1950s were discovered in Izmailovo, Brateyevo and Leninsky Prospects in the heart of the city. They would like, if in seismic zones, to live in earthquake-proof or at least anti-seismic houses, whereas in, e.g., Armenia, Kamchatka and Kazakhstan ordinary houses are built. They would like to stop all-union projects to 'remake' nature which are authorised, elaborated and launched many hundreds or indeed thousands of kilometres from the places of their implementation. They would like to have environmental legislation giving opportunities to protect the individual's 'ecological safety' in court.

Besides all this, they would like to put the military-industrial complex, accounting for more than 90 per cent of the USSR's environmental pollution and overall waste and ravage, under some restriction and control. The whole country is covered with smoking chimneys and waste discharge pipes of 'secret' enterprises (officially called 'postboxes') which are totally uncontrolled environmentally. The military–industrial complex has the right to consume as much material and natural resources as its needs; it has established 'prohibited zones' throughout the USSR and installed about 20 active nuclear Chernobyl-type power units producing plutonium for warheads; it is still producing far more weapons every year than the USA.

The perestroika process can be regarded as very significant not only in a domestic but also in an international ecological context. It should, however, be noted that perestroika has failed to eliminate the environmentally dangerous system of centralised management and decision-making which is capable of initiating such catastrophes as Chernobyl and the shrinking of the Aral Sea. Moreover, most of this centralised system is being preserved because perestroika itself has been designed as an all-union centralised project. This 'revolution from above' includes, for instance, the programme of conversion of the USSR's military-industrial complex to peaceful production which requires the preservation of powerful mechanisms of centralised management. The decentralisation process in the era of perestroika is thus very insignificant and should not be over-estimated. For the traditional all-union centralisation is, after all, being replaced by the same pattern of centralisation at a republican level (the Russian Federation, the Ukraine, Kazakhstan, etc.) with the same centralised pattern of decision-making mechanisms and procedures. In addition, according to interviews with the ministers of the USSR's all-union government published in *Pravitel'stvennyi Vestnik*,[10] practically all the several dozen ministers and ministries (with very few exceptions probably including the Ministry of Culture) are working 'for the defence of the state' just as they did in the years before perestroika.

Thus, it seems possible to conclude that perestroika and glasnost have failed to modify the destructive nature and monopoly of the state political-economic mechanism and failed (so far) to create new democratic tools for environmental protection and resource management. And one should not forget that environmentally dangerous economic-political systems also exist in China, Vietnam and North Korea.

These are the so-called 'preservational' aspects of perestroika, in other words, a relative success of the old decision-making mechanism and management system to preserve itself in this 'period of reforms'.

However, there is another aspect of perestroika more and more obvious of late, which can be called its 'destructive' effect. Undoubtedly the centralised system of decision-making and management is dangerous environmentally, which is confirmed by seven decades of building socialism and communism in the USSR. But this ineffective system had been functioning none the less and was fairly successful in building concentration camps, heavy industries producing severe pollution, railways in the tundra and taiga, canals, hydro-power stations, nuclear military sites and power plants including Chernobyl and the 'sarcophagus' after the disaster, etc.

What is happening nowadays in the period of perestroika? Side by side with dismantling the communist system a destruction of economic links is taking place both in the USSR and Eastern Europe. 'The International Geographic Division of Labour' in COMECON known to every Soviet schoolchild, does not exist any more. And COMECON itself is about to disintegrate. No one has to observe economic obligations any more either in the former 'socialist camp' as a unit or inside any individual erstwhile socialist country. While in the pre-perestroika period the functioning of the Soviet state-owned economy was somehow regulated by the communist party bureaucracy's orders and control, today, with the development of perestroika and the abolition of the CPSU's constitutionally fixed leading role, other (particularly, market) economy regulators are lacking. And, thus, the situation turns out to be catastrophical. The administrative-command economy does not function any more, while a market economy has not yet appeared (and for many political reasons is not likely to be brought into being). The monetary system, even though unsatisfactory in pre-perestroika times, is now almost destroyed. The consumer market has completely fallen to pieces too. The break in economic links between various branches of the economy has evoked breaches in the intersectoral balance and paralysed many industries which are dependent on component suppliers.

This situation is aggravated by the paralysis of all-union power when there are republics, regions and sometimes even city districts proclaiming their sovereignty, ending their subordination to the central power and considering themselves no longer bound by any agreements or obligations.

Sadly, perhaps one must admit that perestroika is likely to increase the destruction of nature, albeit largely destroyed before. After all, perestroika has failed to create mechanisms to regulate the economy or prevent either the former centralisation turning into totalitarianism or a newly emerging democracy turning into chaos. Today, in the era of the communist system's disintegration, there is frankly no room for confidence that if a nuclear disaster similar to Chernobyl were to occur, it would be possible to build a similar protective 'sarcophagus'. The old system is being destroyed; a new one does not yet exist.

Unfortunately, none of the alternative large-scale programmes of economic reform (neither the Leonid Abalkin government programme, nor the popular Grigory Yavlinsky's '500 days' programme turned down by the Soviet parliament) presupposes any changes for the better of the USSR's ecological situation. On the contrary, these are both accentuated by anti-ecological programmes and are obviously intended to put off the solutions to environmental problems for an indefinite period and, indeed, disregard them for the sake of achieving short-term or instantaneous goals. This is a fairly typical approach of perestroika leaders to environmental problems if the problems are in any way within a current political context. For instance, Gorbachev, addressing Soviet *arendatory*, the new category of individual farmers, advised them to pay no attention to the question of water pollution while increasing their numbers of cattle. Or, in Lithuania, the Sajudis popular front leaders who had made their names and gained the population's respect in fighting the all-union Ignalina nuclear power station project, today in the conditions of Lithuania's fight for independence and Moscow's economic blockade, are thanking God for having just this nuclear-power station.

The realities of life are definitely determining people's consciousness at this time. They are in no mood for thinking about environmental problems. They are in no mood for thinking about salmonella in eggs because there are no eggs in shops any more.

Unfortunately, in the period of perestroika and a worsening economic situation the state's primary goals – to feed people and to achieve economic stability – are again 'justifying' the convenient means of their achievement.

It could be very dangerous environmentally if the political situation is destabilised in a state without a proper rule of law, full of nuclear and chemical weapons and operating Chernobyl-type nuclear power stations. In one radio broadcast in the autumn of 1990, Vladimir Matusevich, a Radio Liberty commentator, called the contemporary USSR the 'Upper Volta with nuclear warheads'. Then, apologising, he admitted that the comparison was not pertinent, for Upper Volta was successful in nourishing its citizens while the USSR was not. The imagery used by Matusevich may be offensive, but it certainly reflects the ecological context of perestroika in the USSR.

The ecological context of Soviet life and the diseases of its system have been studied to a greater or lesser extent. But what do we know about the potential after-effects, environmentally speaking, of the disintegration and death of the Soviet system?

Directions of eco-sovietological research

One of the most important and fundamental directions for eco-sovietological research could be the historical analysis of the Marxist-Leninist approach to the environment and resource management: the views, statements, decisions, and so on of Marx, Engels, Lenin, Stalin, Khrushchev, Brezhnev, Gorbachev, etc. The study of environment and resource management history in the USSR, particularly its policy since 1917, would be also useful. Such research has already been started by us under the auspices of the Independent Ecologists' Foundation and includes the following aspects:

1 an analysis and review of the ideology, laws, Communist Party and government statements and strategic economic decisions which have affected the state of the environment in the USSR and on the global scale; the study and assessment of the environmental consequences of the CPSU party 'lines' towards industrialisation, the collectivisation of agriculture, the 'cultural revolution' of the 1920s and 30s, development of virgin land, 'chemicalisation', land amelioration, etc.;
2 the economic evaluation of environmental damage from 1917 to the present;
3 the dynamics of the state of the environment and natural resources in the USSR from 1917 to the present.

The recent history of the 1980s and the perestroika period should be considered within the context of the entire period of Communist rule. In the 1980s the USSR had four General Secretaries of the CPSU, each of whom had his own specific policy of economic and resource management. During this period the USSR experienced drastic changes in its economic policies which directly affected the environmental sphere (for example, the 1984 Chernenko line of developing land irrigation and amelioration).

The comparative analysis of different periods of the USSR's history could include consideration of the following:
1 the state of the environment: air and water pollution (including trans-

boundary pollution and pollution of the seas), land, forests, ecosystems, public health, technology, military impacts, etc.;
2. resource consumption of the economy: estimates of land, water, timber and consumption of other resources based on per unit production or per person of population;
3. the Communist party, parliament and government decisions (projects) and their environmental impact on international, national and local levels; the analysis of chains of causes and environmental consequences;
4. ecological expertise (environmental impact assessments of new projects and strategies);
5. major environmental disasters;
6. major environmental improvements (if any);
7. the state of environmental information and statistics, coverage of environmental issues in the national and local mass media ('ecological glasnost');
8. public environmental movements and their political and environmental role.

Special attention should be attached to Gorbachev's period of reforms (1985–) with its rapid changes in all aspects of Soviet life: the decentralisation and transfer of power to the Soviets; less secrecy and more glasnost, and more activity in public environmental and national movements; and at the same time rapid economic collapse, national tensions and the disintegration of the country. One must take into account that in the era of perestroika the USSR's environmental future seems uncertain since the primary aspects of reforms are still under discussion; one must also consider that reforms can stop at any moment or somehow be reversed as has happened in China and now indeed seems to be happening in the USSR. An assessment and ecological forecast of perestroika's after-effects is thus one of the principal and urgent tasks of eco-sovietology.

How much collapse of the USSR is environmentally safe for the rest of the world and, as we have indicated, even for mankind's survival? It is eco-sovietologists who might best answer this crucial question.

Notes

1. The author of this book, Zeev Wolfson (written under the name of Boris Komarov) is now the editor of the *Environmental Policy Review: The Soviet Union and Eastern Europe*, Jerusalem. See list of contributors.

2. *The Times*, London, 9 February 1991. See also note 8.
3. Mikhail Lemeshev, 'The destructive steps of "acceleration"' (in Russian), *Moskva*, no. 6, 1990.
4. *Sovetskaya kul'tura*, 4 August 1990.
5. *Sostoyaniye prirodnoi sredy v SSSR v 1988 g.* (The State of the Natural Environment in the USSR in 1988) (Moscow: Goskompriroda, 1989) (in Russian).
6. *Okhrana okruzhayushchei sredy i ratsional'noye ispol'zovaniye priroduykh resursov v SSSR: statisticheskii sbornik* (Environmental Protection and Rational Nature Management in the USSR), Statistical abstract of the USSR State Committee on Statistics (Moscow: Finansy i Statistika, 1989).
7. A more detailed review of the Goskompriroda report and the Goskomstat statistical abstract (by I. Altshuler and Yu. Golubchikov) is published in *Environmental Policy Review*, 4, no. 2 (1990), and contains a reference to Nikolai Vorontsov's television address to the nation. Unfortunately, the word 'improvement' was printed in error instead of 'deterioration'. Vorontsov's actual words were 'We must switch from a sharp *downfall* to a slow *deterioration* of the situation'.
8. Figures taken from the report of A. Yablokov, a prominent Soviet ecologist, 'State of the natural environment in the USSR', in manuscript form only, 1989.
9. See *Dritter Internationaler KFK/TNO Kongress Über Atlastensanierung, Karlsruhe, 10–14 December 1990. Late contributions* (by R. Mnatsakanyan, p. 2).
10. Official newspaper of the USSR Council of Ministers.

13 Environmental issues in the Soviet Arctic and the fate of northern natives

Alexei Yu. Roginko

For decades, Soviet economic policy and practice in the use of Arctic resources and spaces were dominated by an approach which can be formulated as follows: 'the more we take from the Arctic, the better'. At different periods this attitude was typical, in our opinion, of other industrialised countries as well. It was an inherent feature of the period of extensive development of the scientific and technological progress in the economic sphere. And when contradictions emerged concerning different uses of the Arctic, or between environmental and social priorities on the one hand, and resources development on the other, they were, as a rule, resolved to the detriment of the former. For a long time all economic activities in the North have been exclusively resource-oriented. Such a situation can be accounted for, if not justified, by our history, by initial low levels of economic development, by low living standards, and so forth. Still, the Soviet Union relies heavily on the Arctic for the supply of fossil fuel resources: the Siberian Arctic accounts for almost two thirds of national oil and more than 60 per cent of natural gas production,[1] and these percentages are bound to increase substantially, especially when Arctic offshore oil and gas development starts. Even now several administrative regions of the Yamal-Nenets Autonomous District produce more oil and gas than the United Arab Emirates.

Such an approach to the development of the Soviet North has resulted in a deep conflict between the economic interests of industrial civilisation and the Arctic ecosystems, now functioning at critical levels. Even more important, it is the interests, identity, and very existence of small northern aboriginal peoples which are now at stake.

Plans for industrial development of the Arctic and subarctic areas have always been met with great anxiety everywhere in the world. Governmental and public organisations usually require from private companies reliable safeguards for the protection of indigenous peoples' needs and interests. These safeguards are also provided for by the ILO (International Labour Organisation) Convention No. 107 for the protection and integration of indigenous and tribal populations to which the Soviet Union is a party. As the experience of other countries suggests, there exists a real possibility to harmonise the interests of native peoples with industrial development.

What is the situation with regard to the protection of northern natives' interests and their human environment in the Soviet Union? The answer is unequivocal – up to the present it has been extremely alarming. These interests were not taken into account in the 1950s, when large-scale nuclear tests were performed in the Arctic. According to the data of the Leningrad Institute of Radiation Hygiene, in the period of atmospheric nuclear testings, total beta-activity in reindeer and lichens has exceeded the natural level ten times. In 1988 the radiation load on the native population was twice the national average and roughly equalled the mean radiation dose received by the population residing in the areas contaminated after the Chernobyl accident.[2] And the northern peoples' interests are not allowed for today while the state is prospecting geologically in tundra and taiga, producing oil and gas, building enormous pipelines across pastures and hunting areas, and so on.[3]

The few improvements that industrial civilisation and technology have brought about in the life of northern natives are far outweighed by the damage inflicted upon the Arctic environment by the ministries, agencies and organisations conducting large-scale, practically unregulated and uncontrolled industrial development. Here are some facts and figures illustrating the current situation.

According to official data, during the past twenty years more than 20 million hectares of reindeer pastures in the North have been destroyed, including about 6 million hectares in the Yamal-Nenets Autonomous District alone. In fact, the situation is even worse, because this figure does not include the so-called 'recultivated lands', which will become suitable for grazing again in fifty years or so provided that they are not used in the meantime.[4] As a result, the reindeer population has decreased from 2.4 million in 1965 to 1.8 million in 1986.[5] According to an assessment made by local scientists from the Yamal Agricultural Station, the damage inflicted on the environment of the Autonomous District is estimated at approximately 60 billion roubles.[6] For comparison, the total national defence budget for 1988 amounted to about 77 billion roubles.

The Soviet North was being 'developed' with outright violations of environmental legislation. Three gigantic natural gas fields in Western Siberia (Medvezhe, Urengoi, and Yamburg) have been projected and developed without any environmental impact assessment. Most valuable species of fish and fur animals have disappeared here completely and dozens of small rivers have been polluted and destroyed. In 1988 a tremendous oil spill from a damaged pipeline occurred in the River Ob' estuary, and its consequences will be felt for decades.

As a result of clumsy, careless amelioration in Magadan Region many rivers and lakes in Chukotka are now devoid of plankton, the feed basis for salmon and other valuable fish. The damage is estimated at 4.7 billion roubles.[7] Fishery resources in many inland northern waters are close to depletion. Intense overfishing accompanied by land-, air-, and vessel-source pollution have undermined fish stocks and trophic chains in the Barents Sea; the very genetic fund of fish and marine mammals here is at risk.[8]

The entire Siberian Arctic coastline, Arctic islands and polar stations are covered with mountains of empty fuel barrels which are not taken away, piles of garbage, scrap, etc. Moreover, according to the data of the Committee of People's Control, the quantity of pipes, construction materials, machinery and equipment abandoned in the North would have provided a reliable material basis for the development of an average-sized European state.[9] In fact, today there is more abandoned scrap in the tundra than there are wild animals. Thousands of square kilometres of man-made Arctic deserts which, taken together could have constituted a 'second Alaska', silently witness a triumph of mindless technocratic approach.

It is the environmental situation in the most industrially developed area of the Soviet North, the Kola Peninsula, which causes the gravest concern. Emissions of sulphur from the peninsula are twice as great as those from the whole of Finland. The so-called industrial desert in this region with practically no living plants covers an area of about 100,000 hectares. The area where sulphur deposition is estimated to be 1–2 grams per square metre annually, amounts to about 5 million hectares, approximately half the size of Finnish Lapland; in that area trees are defoliated and changes in the composition of lichen and moss species are observed.[10] The boreal forest line here is gradually moving to the south, and if this trend is not curbed the peninsula will turn into a rocky tundra in a few decades. The same is true also of the Norilsk area. Up to 60 per cent of the local Kola population suffers from respiratory and other environment-connected diseases.[11] The town of Apatity ranks first in the country by the number of kidney and liver disorders per thousand of inhabitants. In other words, the Kola Peninsula has already turned into a real environmental disaster area.

Effects of the North's degradation

No wonder that such an environmental aggression sometimes encounters spontaneous opposition and protests from the indigenous population of the North. For instance, in the valley of the River Sob' (tributary of the Ob'), where the small Khanty people live, several areas have for centuries been considered sacred; it was strictly forbidden to fish and to hunt there, to cut forest and to make fires. That was an aboriginal method of protecting fish spawning areas and waterfowl's nesting places. One can imagine the Khanty's indignation when a few years ago powerful dredges arrived and started to mine for sand and gravel there, actually destroying the river. So the Khanty loaded themselves and their families into their small boats and blocked the way for huge steel vessels.[12] They have won in this case, and their victory proved for the first time the possibility of collective self-defence. But of course such methods could not provide for a comprehensive solution of the whole problem.

Reckless, aggressive exploitation of the northern environment by the Soviet industrial ministries undermines the natural basis of small indigenous peoples' existence. Their human environment, traditional lifestyles, material culture and social organisation are being changed so drastically that it is difficult to guarantee their survival in the coming decades. For example, forced resettlement of northern natives has led to the organisation of large sovkhozy, usually operating at a loss, replacing formerly quite profitable small farms. Uncontrolled overgrazing in their vicinities has undermined the feed basis for reindeer husbandry. Being isolated from traditional trade businesses and the environment to which they have been accustomed, the tundra natives have considerably reduced their settlement and hunting areas and shortened their migration routes. As a result, they are gradually losing their traditions of reindeer husbandry, hunting and fishing. When living in small towns, as a rule they occupy the lowest rungs of the social ladder and the consequences of traditional life-style destruction are unemployment, high criminal and suicide rates, alcoholism, and so on. Average life expectancy of many small northern peoples is comparable only to that in the least developed countries – 43–45 years for men, about 55 years for women. Infant mortality has also been exceptionally high.[13] And this is not an effect of a certain universal process dooming any small peoples to extinction, as some would seem to suggest, but rather a direct result of an incompetent social policy. To cite just one fact, while back in the 1960s average life expectancies of North American Inuits and Soviet northern natives were roughly equal (62 years), two decades later the same index has increased by 10 years on the one side of the Bering Strait and decreased by 10–15 years on the other.[14]

The question might arise whether we really need that untouched northern environment, those northern natives' traditional life-styles preserved? After all, they number only 184,000 – about 1.5 per cent of the total population of the area. The author's point of view (shared by a number of Soviet researchers) is that besides humanistic, ethical considerations, which are in themselves more than sufficient, there exists a purely pragmatic reason. No matter how paradoxical it might sound, the preservation of the Arctic environment, the traditional life-styles of tundra natives and their ancient culture is an essential prerequisite for the successful industrial development of vast northern territories. Without careful and thoughtful study and the use of centuries-long aboriginal expertise we will not be able to exploit rationally the natural resources of the Arctic or establish a sound food basis for its development. Instead of extinguishing a nomadic way of reindeer husbandry we should support and develop it (as is being done in Finnish Lapland), for it has been proved that it is the most rational way of using fragile tundra landscapes without destroying them.

The roots of the northern problem

What are the roots, the principal causes of the current conflict, even crisis, in the Soviet North?

First, there is the property issue. The resources of the northern rivers, forests and tundra as well as the northern lands and waters themselves have long ago ceased to be a collective property of the native population. They have been passed over to the government, to the state, and the state property, especially in the North, has long ago become a fiction. It is the ministries' and the agencies' property that exists in reality. And as these organisations are guided primarily by their selfish, narrow-minded interests, they are not even trying to coordinate their activities with the urgent, vital needs of the northern natives. Seizing lands and waters without their real owners' permission, the ministries just pay compensation to each other in case of environmental damage. And it is only the people, the northern natives, that have been forgotten. Their living standards are extremely low: in their settlements more often than not there are no schools, hospitals or electricity, let alone such modern conveniences as water supply and sewerage. Their average wages are 10–15 times lower than those of the oilmen working and living nearby.[15] And that is happening in the region that supplies a large part of Europe with energy resources.

The second cause is insufficient, scanty knowledge about the state of the Arctic environment. One of the important reasons for this lack of information is that the study of the North has been fragmented into a number of separate branches. Analytical approaches have prevailed and, moreover, a type of scientist has been created who is ready to carry out any orders of the financing agency. Environmental research in the Norilsk area is financed by the Non-ferrous Metals Ministry, and that in Western Siberia by the Ministry of Oil and Gas Construction. Needless to say, those environmental protection departments which are functioning within the ministries' structure serve more to cloud and conceal the real situation than to clarify it.[16]

Thirdly, there is a lack of finance, which is really a national rather than a regional problem. About 10 billion roubles are spent on environmental protection all over the country annually compared with 80 billion dollars in the USA. But this problem is particularly acute in the North, where sums allocated for the prevention of environmental damage and the restoration of environment where it has been disturbed are not comparable to the damage already inflicted. For instance, out of 32 billion roubles initially allocated for the development of natural gas fields in the Yamal Peninsula, only 300 million roubles (i.e. less than 1 per cent of the total project cost) was to be spent on 'compensatory measures', including environmental protection. For comparison, environmental outlays during construction of the Alaska pipeline have accounted for about 15 per cent of the total project cost. In Yamburg, where natural gas production is already under way, the Ministry of Oil and Gas Construction plans to spend some 150 million roubles for the restoration of the environment and for damage compensation, while the real cost of environmental damage here is in the order of billions, not millions.[17]

The final cause, which is really a result of all the factors enumerated above, is the absence of a comprehensive strategy for the Arctic or clear concept of sustainable development in the North which could have balanced economic and environmental concerns.

What is being done in the Soviet Union to ameliorate the existing situation in the North and to resolve this current environmental conflict? First, the seriousness and the urgency of the issue are now fully realised, and several measures are being undertaken. They may be in the right direction but are still deficient in several respects, i.e. insufficient, incomplete and uncoordinated.

Good laws are being passed, but their provisions mainly remain on paper, and are not enforced. For example, in November 1984 an 'Ukaz' (Edict) of the Presidium of the Supreme Soviet was passed, entitled 'On the strengthening of nature protection in the regions of the Far North and sea areas adjacent to the northern coast of the USSR'. It envisages the establishment of a network of nature reserves, places strict limitations on transport use, tourism and industrial development in the Arctic, provides for the special design, equipment and manning standards for vessels operating in Arctic waters and for the closure to navigation of specific sea areas at certain times, etc.[18] But the problem is that the concrete norms and rules upon which the execution and enforcement of the edict depend are still not elaborated. Until now, it was once again the ministries involved that have been blocking the issue. Thus, good intentions have not been translated into action and still remain mainly on paper.

Nevertheless, there have been some positive developments. According to the official data, the load of suspended solids and detergents in the Arctic water basins are now half the 1980 figure; oil and oil product loads have also decreased considerably.[19]

One more example to illustrate a more critical and more demanding approach. In December 1987 an arbitrary decision was made by the Deputy Chairman of the Council of Ministers to start construction works on the Yamal Peninsula both to develop one of the largest natural gas fields in Siberia and build a gas pipeline to Europe. This was done practically without any environmental impact assessment and despite the fact that many technical and engineering decisions had not undergone experimental testing. What would have happened earlier, in the so-called 'period of stagnation'? Most probably, the project would have proceeded as planned, and the environmental effects might have been unpredictable. But these days, since the public and newly established State Committee for Environmental Protection voiced their opposition to the project, independent environmental impact assessment was conducted, and the Yamal gas development was suspended due to the 'inadequate technical, economic, and environmental grounds of the project'.[20]

That is of course a positive example, but it can hardly serve as a method for a comprehensive solution of the whole problem, including the issue of preserving the northern natives' identity and their ancient culture.

What is to be done?

So what has to be done, and done immediately? Undoubtedly, a new economic mechanism for environmental management and resource use needs to be developed as soon as possible. Several experts have suggested, for example, developing a new system for assessing land value in the Arctic which would take into account possible environmental damage costs. Of course, more effective legal measures not only for a compensation of such damage costs but for the prevention of damage itself should be elaborated. And, probably, as is suggested by Academician V. Kotlyakov, until such measures are developed, it might be better to suspend temporarily all the industrial development in the Arctic and concentrate efforts on the development of central regions.[21]

The most important thing to realise is that no comprehensive solution can be achieved without the active involvement of the northern natives themselves. Any attempts to implement any, even the most helpful, measures from 'above' – from Moscow or Tyumen', from Magadan or Krasnoyarsk – are doomed to failure. What the central authorities should do is to curb the expansion of industrial ministries to the North and make them respect and consider the needs and interests of aboriginal peoples.

A specific national policy is required to assist the northern native peoples strive for their survival and the preservation of their ethnic identity. And it means more than just to provide the entire population of the North with 'equal rights and equal opportunities'. Under those 'equal' conditions the strongest will always win – and the northern natives are not yet the strongest in their native land. The independence of their development is the only possible means of their survival, because if the barrier of social passivity and alienation is not broken by the natives themselves, no support from outside would help. Active participation of the northern natives in regional and local development programmes on all their stages – from design to fulfilment – should be recognised as a major political principle.[22] It should be for the natives themselves to decide what is better for them – traditionalism or industrial development, reindeer or oil, privileges from the state or economic prospects.

One of the most commonly proposed ways of achieving this aim is to organise in the North a network of reserves with exclusive rights of ownership and disposal of their land, waters and resources held by the native population.[23] This has already been done in the USA and Canada, where about 25 per cent of the northern territories have been turned into biosphere reserves. And it inspires hope that the Communist Party Platform on National Policy adopted by the Central Committee plenary meeting in September 1989 envisages granting to the Soviets of the territories where the northern natives live *'exclusive rights* for their economic development, that is for their hunting areas, pastures, inland and offshore waters, forests, as well as the right to establish natural reserves with the aim of restoring and preserving these peoples' areas of settlement'.[24] At the end of March 1990, the First Congress of Small Peoples of the North was held, and the Association of the Small Peoples of the North of the Soviet Union was established. Its aims are the promotion of the political, social and economic rights of the northern natives, preservation of their cultural identity, control over resource exploitation in the territories of their residence, as well as the representation of these peoples' interests at all governmental levels. In the declaration adopted by the congress, its participants called for a revision of the principles of northern territories' industrial development; they demanded, in particular, that any large-scale project concerning utilisation of natural resources should be examined by the relevant regional native peoples' associations.[25]

The realisation of the measures envisaged will pave the way for the establishment of real self-government structures for the northern natives and will enable them to feel once more the masters of their own lands and waters, tundra pastures and reindeer herds. Only economic self-government, bringing an opportunity to dispose of cooperative property in northern communities independently, will be able to return to the natives both personal and social *raison d'etre*. That is exactly what they need to assist them in their strivings towards self-preservation and cultural identity.

Notes

1. A. Granberg, 'Siberian economy – goals of structural policy', *Kommunist*, no. 2 (1988), p. 32 [in Russian].
2. *Izvestiya*, no. 241, 21 August 1990.
3. A. Pika and B. Prokhorov, 'Large problems of small peoples', *Kommunist*, no. 16 (1988), p. 78 [in Russian].
4. V. Sangi, 'For the top of the crown not to fall off . . .', *Literaturnaya Gazeta*, no. 7 (15 February 1989), pp. 1, 7 [in Russian].
5. Pika and Prokhorov, 'Large problems of small peoples', p. 78.
6. F. Sizyi, 'The price of Yamal', *Ogonyok*, no. 46 (1988), p. 20 [in Russian].
7. R. Bikmuhametov, 'Man-made swamps', *Energiya: ekonomika, tekhnika, ekologiya*, no. 7 (1988), pp. 20–4 [in Russian].
8. V. Smirnov-Semyonov, 'The Barents trouble', *Izvestiya*, no. 241 (28 August 1989) [in Russian].
9. Sizyi, 'The price of Yamal', p. 21.
10. Martti Varmola, 'The state of forests in Finnish Lapland', a background paper prepared for the Consultative Meeting on the Protection of the Arctic Environment, Rovaniemi, 20–26 September 1989, p. 2.

11. V. Kiselev, 'What is ahead, Monchegorsk?', *Sovetskaya Kultura*, 14 January 1989 [in Russian].
12. L. Shinkarev, 'Tundra: how to help northern natives to preserve ethnic culture', *Izvestiya*, 15 June 1989 [in Russian].
13. Yu. Golubchikov, 'We're losing a second Alaska', *Sovetskaya Kultura*, 13 July 1989 [in Russian]; Pika and Prokhorov, 'Large problems of small peoples', p. 80.
14. V. Sangi, 'To return rights to the owners of the land', *Izvestiya*, 12 July 1990 [in Russian].
15. Pika and Prokhorov, 'Large problems of small peoples', p. 77.
16. Golubchikov, 'We're losing a second Alaska'.
17. See an interview by I. I. Mazur, the Deputy Minister for Oil and Gas Construction, in *Energiya: ekonomika, tekhnika, ekologiya*, no. 3 (1989), pp. 20–5 [in Russian].
18. *Vedomosti verkhovnogo soveta SSSR*, no. 48 (1984), p. 863.
19. Consultative Meeting on the Protection of the Arctic environment, Rovaniemi, 20–26 September 1989. Statement of the Soviet Delegation, p. 4.
20. For details see: Sizyi, 'The price of Yamal'; V. Larin, 'Aral . . . Baikal . . . Yamal?', *Energiya: ekonomika, tekhnika, ekologiya*, no. 3 (1989), pp. 18–20 [in Russian]; V. Kalyakin, 'A departmental variant for Yamal', *Kommunist*, no. 5 (1989), p. 68 [in Russian].
21. See V. Kotlyakov and G. Agranat, 'Tropics of the North', *Pravda*, 9 May 1989 [in Russian].
22. See Pika and Prokhorov, 'Large problems of small peoples', p. 82.
23. See, for example: L. Shinkarev, 'Tundra: how to help northern natives . . .'; Sangi, 'For the top of the crown not to fall off . . .'.
24. 'National policy of the party in the current situation (the CPSU Platform)'. Adopted by the CPSU Central Committee Plenary meeting on September 20, 1989, *Pravda*, 24 September 1989 [in Russian].
25. *Izvestiya*, no. 92, 1 April 1990.

14 Air and water problems beyond the Urals

John Massey Stewart

Three centuries ago the Russians conquered Siberia in little more than sixty years. Our own century has seen the Russian 'conquest of nature' in the same huge area within much the same period of time. It has been on a massive scale not so much due to its continental size but to Stalin's army of slave labour and the rushed industrialisation, the command economy, policy of permanent settlement and, most basic of all, the enormous wealth of the natural resources: minerals, water power, oil and natural gas, endless forests.

However, the 'conquest of nature' is surely a contradiction in terms. Nature cannot be conquered, only disrupted to a greater or lesser extent. Here the extent is emphatically greater and, both because of the magnitude of the Soviet enterprises and the fragility of northern ecosystems – tundra and taiga usually underlain by permafrost – nature has taken its revenge. The environment suffers badly in all too many cases, species of flora and fauna are at risk (if nothing worse), and *Homo sapiens* is subject or potentially subject in great numbers to serious environment-related health problems including high infant mortality and a shortened life span. The infant mortality rate nationwide is indeed 2.4–5 times the rate in the USA, France, Britain, the former West Germany or Japan.[1]

Since World War II the largest constructional project east of the Urals has been the building of the BAM railway more than 4,300 kilometres from Baikal to the Amur River across twenty-two mountain ranges and seventeen major rivers. Since the constructional zone was to encompass twenty-five million acres of forest, a major environmental impact was inevitable, and scientific commissions, sensibly appointed to report on the problem, warned that nature in the area was 'very vulnerable. Hence we must treat it carefully and cautiously'.[2]

Despite much environmental research beforehand, a special council for nature protection, and recommendations, prohibitions, and many conservation regulations (perhaps just on paper?) during its construction, much environmental degradation resulted given the scale of the project and sensitivity of the ecosystem. Large areas of taiga were felled (some unnecessarily), moss cover was damaged and the exposed permafrost areas thawed, leading to swamps, landslides, subsidence and erosion. Gravel extraction from river beds for the construction of embankments, totalling 250 million cubic metres, ruined many rivers. By 1984 thermo-erosion was found practically the whole length of the railway, and innumerable gullies of between 3 and 300 metres deep have formed across an area of more than 1 million square kilometres.[3] Many of the towns and settlements sited conveniently in the natural depressions along the railway suffer air pollution not only from the new local industries but from inversion, trapping the fumes of vehicle engines left running, sometimes overnight, in the extreme winter for fear of not being able to start them again (although, granted, this is not a specifically BAM problem).

The most fragile ecosystem of all is that of the tundra which regenerates extremely slowly and sometimes not at all. In winter the snow may protect the thin layer of vegetation from tracked and heavy vehicles, but with no snow cover 'a vehicle has only to cross the frozen tundra once to destroy the scanty covering of the permanently frozen soil. The permafrost starts to thaw, pits and swamps form, and it takes decades for the scars to heal.'[4] To this delicate environment has come the whole infrastructure of mineral extraction with all its equipment and personnel, often leading to an irreversible process of degradation. Not only have the pastures of the world's greatest reindeer population, both wild and domesticated, suffered, but many streams and rivers have been polluted, particularly by prospecting and extraction. 'All along the northern coast', reported Academician Abel Aganbegyan, 'pure fresh water is almost as scarce as housing . . . Yet we heard nothing about any serious or systematic effort to solve this existing problem.'[5]

Aganbegyan's visit was a whole decade ago. Since then, the activity has continued. The Yamal Peninsula is perhaps the worst case; in 1988 the head of the Institute for Problems of the North, Vladimir Mel'nikov, reported that industrial development there had already affected birds' nesting places, twenty-eight rivers rich in fish and six million hectares of reindeer pasture. 'If one sticks to the old approach – gas and oil at any cost', he warned, 'then the Arctic tundra will be in for an ecological catastrophe.'[6]

Public opinion on the peninsula has been becoming increasingly restive. In 1988 the Deputy Minister of Construction of Oil and Gas Industry Enterprises visited the Yamal's boom town of Nadym and was heckled by local ecology activists, angered by the Politburo's decision to proceed with the peninsula's gas exploration without waiting for environmentalists to finish their research.[7] The following year the Praesidium of the USSR Council of Ministers decided that the development of the Yamal Peninsula should be postponed. For the reindeer breeders, fishermen and hunters it had been a worsening situation, increasingly deprived of their means of existence: their fish disappearing from polluted rivers and lakes; their reindeer's Iceland moss fodder often destroyed by man's disturbance; their game animals retreating from his noise.[8]

Air pollution

The tundra is under stress not only on the ground but from the air as well. Acid rain, indeed, now affects the whole tundra from the Yamal Peninsula as far east as Kolyma, excepting only the eastern side of the Taimyr Peninsula, so that the lichen and Iceland moss, the reindeer's staple foodstuffs, are being destroyed, and the herds, both wild and domesticated, are declining.

Air pollution is probably the gravest environmental problem east of the Urals because of its concentration in urban areas and the millions of lives thus affected. The problem is scarcely confined to Siberia and the Soviet Far East – or indeed to the Soviet Union – but the territory provides some of the worst examples in the country. In Chelyabinsk, for instance, on the eastern side of the Urals, and other industrial towns of the area, 'life is a living hell . . . [with] clouds of noxious smoke to breathe . . . a life expectancy five to eight years below the national average . . .' and a cancer rate for Chelyabinsk oblast of 291 (compared to the national average of 129) per 100,000.[9]

Unfortunately, this does not seem an exceptional case. Official 1988 statistics list sixty-eight Soviet cities with a total population of forty million which are becoming increasingly dangerous to live in.[10] Chelyabinsk, not technically in Siberia although east of the Urals, is included in this list as are eighteen cities of Siberia proper and the Soviet Far East. Toxic emissions from four of these are here compared (using data from another table) with those of Moscow and the worst case of all, Krivoi Rog in the Ukraine.

Fortunately, the cities cited (except for Novosibirsk) show a decrease of emissions (although actual toxicity could have risen). According to this selective table, Novokuznetsk, a city of 600,000 (1989) in the Kuzbass, has the USSR's worst air-pollution after Krivoi Rog. 'On every side', according to a recent visitor,

mile upon mile of chimneys belch grey, brown and black plumes into the pale northern sky. On a clear day the smoke from the West Siberian Metallurgical

Table 14.1. *Toxic emissions into the air from stationary sources for selected cities (in thousands of metric tons)*

	1985	1987	1988
Barnaul	208.5	183.7	183.6
Krasnoyarsk	341.8	291.0	258.6
Krivoi Rog	1,314.2	1,290.0	1,252.7
Moscow	411.0	369.1	311.8
Novokuznetsk	1,001.9	892.9	833.0
Novosibirsk	232.1	228.4	235.2
Total USSR	68,344.9	64,295.7	61,716.2

* Source: *Ekonomika i zhizu'*, no. 4 (January 1990), p. 18, via CDSP, 28 February 1990, vol. XLII, no. 4 (1990).

Kombinat, the largest metal works east of the Urals, can be seen 60 miles away. According to the magazine *Nauka i Zhizn'* (Science and Life), every year 833,000 tons of effluent, three quarters of it carbon monoxide and sulphur dioxide, responsible for acid rain, shower upon its 680,000 inhabitants. In the outlying area of Mezhdurechensk, strip mining of 20m tonnes of coal a year has turned swathes of virgin taiga forest into moonscapes.[11]

None of the city's factories evidently has proper cleaning equipment but, as so often, there is no money to eliminate pollution. Life expectancy in Novokuznetsk is 8–10 years below the national average, the incidence of eye and respiratory disease in children and malignant tumours exceeds the all-union level, and there is a very high incidence of cancer, bronchial disease and heart problems.[12]

Oddly enough, Noril'sk, on the Taimyr Peninsula, is omitted from the list of the sixty-eight worst-polluted towns. The largest town in the world north of the Arctic Circle (pop. 174,000, 1989), Noril'sk produces some two-thirds of total Soviet nickel production (as well as many other strategic metals), officially emitting about 2.5 million tonnes of sulphur dioxide a year, a figure above municipal norms and much the same as for the whole of Canada. The development of the copper nickel sulphide deposits nearby has produced 'an arctic sky perpetually stained by a sulphurous yellow smog',[13] and unfortunately for the citizens, due to local geological and topographical factors, the residential area was built perforce downwind of the factory chimneys,[14] with a high rate of respiratory and other diseases in consequence. The Noril'sk mining and ore-dressing works has been accused of the death of one and a third million acres of taiga – an area the size of Norfolk – and ordered to pay the huge fine of 27 million roubles, the estimated cost of the environmental damage.[15]

Not surprisingly, a local green movement has emerged (as in Novokuznetsk). The (so far tiny) Taimyr Green Front challenges official pollution level figures which it sees as grossly underestimated and claims the local authorities do not fulfil the Central Committee's call for a reduction of discharges into water. (Curiously, air is not mentioned.) In a press release the front has declared,

Northern industrial concerns using outmoded technology have been transformed into 'factories of death'. Millions of hectares of northern forests have perished; fish have disappeared from murky waterways; deer have been decimated by fatal maladies. The remaining representatives of some [small northern] nationalities are poised on the edge of extinction.[16]

Canadian scientists believe that the great Noril'sk complex contributes markedly to Arctic haze, acid rain and marine pollution (via the Yenisei River);[17] and their American counterparts have identified suspended particles in central Alaska, across the Arctic Ocean, as consistent with nickel and other heavy metals from Noril'sk.[18] If this is so, it is not just a trans-boundary, but a trans-oceanic problem.

The foreign visitor beyond the Urals finds air pollution all too evident with chimneys emitting grey or black smoke by day and often night and sometimes horizontal smoke-ribbons or bands of smog up to several kilometres long clearly visible on windless days. According to one children's hospital in Ulan-Ude, the main health problems for its young patients are respiratory, due firstly to Buryatia's harsh winters and secondly to the city's air pollution. Yet there seems less concern about Ulan-Ude's air pollution than about the state of Lake Baikal, seventy-five kilometres away. However, west of the lake in the autumn of 1989 a class of Irkutsk schoolchildren and their teacher demonstrated against a factory chimney built so short and wide that pollutants were falling onto homes in the immediate vicinity.[19]

Water problems

This huge area east of the Ural Mountains also has its water problems. For a start, West Siberia's vast deposits of oil and gas have produced much welcome foreign currency in exports to the West but large-scale environmental problems as well. Giant casing-head gas flares have been burning for a quarter of a century,[20] creating a virtually dead zone for some distance. Oil leaks from burst pipelines due, for instance, to faulty welding have been estimated at 1 million tonnes a year.[21] And since West Siberia's enormous wetlands and innumerable lakes make it an area of almost more water than land (particularly during the thaw) these oil spills can pollute extremely extensive areas.

There are many cases in general of deteriorating water quality, pollution of reservoir water and poisoning of fish and waterfowl stocks.[22] The timber industry has been a leading culprit. Billions of roubles worth of felled timber lie beneath Siberian rivers, poisoning the water, killing the fish and preventing spawning. And clear-cut slopes have brought erosion and the ruin of many rivers.

The great post-war hydro-electric projects have also caused environmental disruption and not just because of the great areas flooded. Ten years ago, in 1982, A. S. Isayev, director of Krasnoyarsk's Sukachov Institute of Timber and Wood of the USSR Academy of Sciences' Siberian Division, complained that

the ecological cost of building a hydro-electric station is ignored. The Bratsk reservoir, for example, drowned millions of cubic metres of prime Angara pine. [Much of it was felled but left on the bed as transport could not be arranged in time.] Millions more cubic metres of timber went under water with the construction of the Ust-Ilimsk hydro-electric station.[23]

At the Sayano-Shushenskoye project, where timber was not felled, whole trees have risen to the surface each summer so that all the inlets, bays, river mouths and the entire area behind the dam's intake have been 'crammed full of floating trees'.[24]

At Novosibirsk, according to Isayev, the reservoir flooded the Ob' region's most fertile land. At Krasnoyarsk the reservoir changed the local climate; for 150 kilometres below the dam (not 20 km as predicted) the Yenisei no longer freezes and the change in its temperature has increased colds, caused fog, and made river pollution harder to combat by substantially reducing the amount of heat the Yenisei discharges into the Arctic Ocean.[25] All these are the big hydro-electric schemes, but even the smaller ones have their effects. At Gusinoozyorsk, for instance, in Buryatia, the temperature of Lake Gusinoye has risen, altering the lake ecology. And the new industries powered by these hydro-electric schemes have often brought air pollution. For many miles around Bratsk, for example, the casual visitor finds trees almost continuously dead or dying.

On the distant Kamchatka Peninsula, the greatest wealth is the Pacific salmon which returns to spawn in the rivers each year. It is already desperately scarce in the Amur, the Primor'ye and on Sakhalin Island. Yet here, virtually the whole length of its biggest river (the Kamchatka) has been despoiled. The 'incompatible and uncontrolled activity' of different governmental departments – the former Agroprom, forestry, mining interests, etc., is blamed as are geological parties, tourists, and the 'logging barbarians' who have destroyed the valley's forest, caused the tributaries to dry up and blocked the main waterway with sunken logs. State farms on the peninsula's river banks have also been destroying Kamchatka's salmon rivers with their irrigation schemes.[26]

On the Chinese border, Lake Khanka is of sufficient scientific interest to be not only a zapovednik or state nature reserve but a biosphere reserve and internationally recognised wetland site under the RAMSAR convention. But it has suffered from much aerial spraying of fertiliser and pesticide on the surrounding rice plantations, resulting in the deaths of many birds and much wildlife. Plans to export the rice were cancelled when it was discovered they contained a dangerous concentration of chemicals.

Nuclear pollution

Nuclear pollution is another problem beyond the Urals. Back in 1957 a huge concrete vessel containing high-level radio-active waste exploded at Kyshtym in the eastern Urals (Zhores Medvedev was the first to reveal the accident). According to an official Soviet report finally released to the International Atomic Energy Agency in 1989, the subsequent fall-out contaminated 15,000 square kilometres of land in the three provinces of Chelyabinsk, Sverdlovsk and Tyumen', necessitating the evacuation of more than 10,000 people.[27] Western Siberian rivers, including the Ob', were contaminated[28] and, thirty-five years later, 'there is some concern that reduced evaporation [in the reservoirs built to stop the heavily polluted local river flow] and the closure of the remaining reactors will cause the water level in the reservoirs to rise, leading to radioactivity being flushed out into the main Siberian river system'.[29]

TASS reported in 1990 that the authorities had shut down one reactor of the Tomsk nuclear power station and would soon close down the other, thereby reducing by half the flow of hazardous materials into the nearby Tom river.[30] In the post-Chernobyl glasnost era, the Soviet media voices much popular disquiet about nuclear hazards within the USSR, both actual and potential, publicising, for instance, that radioactive waste containers were buried for up to fifteen years only 20–30 metres from the Irkutsk Polytechnical Institute.[31] In the Soviet Far East there has been great concern about the possible construction of a nuclear power plant on the Amgun river near Komsomol'sk-na-Amure,[32] and in 1989 authorities in the ports of Magadan, Nakhodka and Petropavlovsk-Kamchatsky announced that they would not allow the nuclear-powered ships of the Far Eastern Shipping Company (FESCO) to berth. This followed the voyage of FESCO's *Sevmorput*, refused entry to the ports of Nakhodka and Magadan and only reluctantly allowed to land its cargo at Vladivostok after protests by tens of thousands of the city's citizens.[33]

According to *Komsomol'skaya Pravda*[34] an action group has been set up in Krasnoyarsk to get rid of what the *New Scientist* has described as 'the world's biggest store for radio-active waste'.[35] After processing, the waste, from CEMA and European nuclear power stations, is scheduled to be deposited in the clay strata 700 metres beneath the Yenisei. Twenty million roubles have been spent so far on piping and concrete, but the issue is as yet unresolved.

One northern minority people, the Nentsy, have been concerned that nuclear tests will be increased in Novaya Zemlya following the decrease of operations in Kazakhstan. In 1989 they called on the Soviet parliament not only to protect the local population from radiation hazards but to outlaw nuclear testing in the region altogether.[36] But there are already Siberians suffering from radiation. The atmospheric nuclear tests of the 1950s and 60s conducted in the Far North have brought tragedy to the native peoples of the Chukotka Peninsula. The Chukchi reindeer herders are the final link in the food chain from lichen via reindeer to man. As a result they have extremely high accumulation of radioactive elements: 10–20 times the amount of lead-210 and 137 times the amount of caesium in their bone tissue than non-reindeer meat eaters. The result is that 90 per cent of Chukchi now suffer from chronic lung disease and almost 100 per cent from tuberculosis; there are permanent outbreaks of virus and bacterial gastro-intestinal infections and parasitic diseases. The cancer incidence in Chukotka is 2–3 times higher and liver cancer 10 times higher than the USSR average, and the death rate from cancer of the oesophagus is now the highest in the world.[37]

Lake Baikal

Probably the most famous environmental case history in the Soviet Union is that of Lake Baikal (albeit now overshadowed by Chernobyl and the Aral Sea).[38] Set in spectacular wilderness scenery, Baikal is revered by Russians and even more so by Buryats, the local native people, who endow it with spiritual attributes. But it is also the world's scientifically most interesting lake, the oldest and deepest with some 1,000 endemic species, and the planet's largest reservoir, holding one-fifth of its unfrozen fresh water, purified by a unique and complex ecosystem. So the feelings that were aroused when Baikal became subject to increasing environmental pressure are understandable.

For some thirty years a long struggle to save the lake has been waged which has only partially been won. The culprits have included the timber industry with its clear felling (in an area of slow regeneration), consequent erosion, silt and landslides, and sunken logs both in Baikal's tributaries (in some cases up to four deep) and the lake itself. At one stage over 130 streams and springs were dry for much of the year and local fish could no longer spawn in 2,200 miles of waterway. Much of this, fortunately, has now been remedied for, as the leading local geographer V. V. Vorob'yov, has written, 'The purity of Lake Baikal's water depends on the state of its basin's forest cover'.[39]

But pesticides and fertilisers from hundreds of agricultural units still leach into the lake, and the basin's manufacturing and extractive industries continue to pollute. In 1989, after decades of debate and resolutions, more than 100 industrial enterprises in the area still had no purification facilities at all, dumping annually millions of tons of waste water into Baikal bearing zinc, mercury, tungsten and molybdenum. And nearly 700 agricultural facilities have contaminated the lake's tributaries with oil and caustic organic chemicals.[40] Baikal's chief pollutor, however, has been the pulp and paper plant built at Baikal'sk on the lake's southern shore and now supporting a local population of some 30,000. The factory began operating in 1966 before even the (substandard) purification facilities were working properly and in the first year and a half of operation alone emitted 383 tons of toxic matter into the lake, creating temporary islands of alkaline sewage, one of them eighteen miles long by three miles wide.

In the face of high-level lobbying by the scientific community, two decrees were passed by the CPSU Central Committee and USSR Council of Ministers in 1969 and 1971, only to be largely ignored. But a third decree in 1977 began to improve matters as did the installation of a sophisticated 46 million rouble purification system. Yet, however pure the effluence, it cannot match Baikal's natural water. By now, two and a half decades after it began to operate, Baikal'sk has poured into the lake more than 1.5 billion cubic metres of effluence. Grigory Galazy, for many years director of the Limnological Institute studying the lake and for long the leading campaigner for Baikal's conservation, has calculated that, because Baikal'sk uses an enormous amount of water every hour, it has now processed 15,000 cubic kilometres of lake water, more than half the total (and once the world's cleanest water), so less than half remains in its natural condition.[41] However, the brutal fact is that the nation is short of paper and Baikal'sk is one of its biggest paper mills.

In 1987 the Central Committee and CPSU Council of Ministers resolved to transfer by 1993 present production to Ust'-Ilimsk 700 kilometres north-west (where, however, there have already been protests about the pollution it will bring) and convert the Baikal'sk plant by 1993 to other use, still, however, undecided. After so many years, the future remains uncertain.

Air pollution is one more problem for the lake. Endangered species of flora in a nature reserve east of the Baikal'sk factory have been suffering from the factory chimneys' emissions and, due to the prevailing westerly winds, the many industries of the Irkutsk oblast west of the lake add to the airborne stress. In 1985, for instance, they contributed 1.2 million tons of air pollutants. Vegetation is already disappearing in some areas.

Baikal's biggest tributary, the Selenga, adds a substantial share of pollution. Already polluted before leaving the Mongolian border to the south, it thus adds an international dimension to the problem. In Ulan-Ude, despite many years of appeals for purification, the river accrues the city's sewage and some fifty factories' industrial effluence, mostly untreated. Further downstream at Selenginsk it adds the inadequately filtered effluence of the Selenginsk Pulp and Cardboard Factory, where the chimneys add their own malodorous contribution to the environment. Here, ironically, giant letters on the outside wall proclaimed on this writer's visit in 1989, 'Comrades, on us lies a special responsibility for the protection of the unrepeatable beauty of Baikal'. In fact, Selenginsk, which began operating in 1978 as a cellulose factory, installed a purification system costing 23 million roubles and is now planning to adopt a closed cycle system with no effluence entering the Selenga or ultimately Baikal.

BAM brought a further problem to Baikal's extreme north-west coast with the construction of the town of Severobaikal'sk. Many of its boiler houses have little or no filtration, and air pollution has become a health hazard with acid rain falling on the town and lake alike. Yet the expansion of its population and industry have been mooted – and in a seismic area to boot. There are also the pressures brought by tourism. Each year some 700,000 visitors camp around the lake, resulting in forest fires, litter and environmental damage.

How much damage has actually been done to the lake by these different factors? For 25–35 square kilometres around the Baikal'sk outlet the water's chemical composition has changed with an alteration of flora and disappearance of some endemic species. Particularly significant is the high mortality rate of epishura – a small crayfish – since it filters Baikal's water and thus plays an important part in its natural purification. The rate of growth and fertility of Baikal's fish and famous freshwater seal have declined and 60–95 per cent of the roe of the largest fish population, spawning in the polluted Selenga, is now destroyed annually.[42] According to Vorob'yov in 1988, 'If the pollution of the air and water basin continues at present rates and in the same manner . . . irreversible changes are inevitable'.[43]

But there are hopeful developments. In 1990 a long term Comprehensive Territorial Environmental Plan (TerKSOP), regarded by environmentalists as very significant for Baikal's salvation, came into force. Amongst other things it is establishing three nature protection zones in Baikal's basin of varying stringency, with maximum protection designated for the coastal belt. And a system of 'environmental passports', i.e. environmental permits for enterprises in the basin, is beginning. Secondly, a law on Baikal's conservation is now under discussion, and legislation should, in theory, be far more effective than the past resolutions which, as resolutions, albeit passed at top level, have lacked teeth. Thirdly is the goal of Baikal's nomination for UNESCO World Heritage Site status which is seen locally as of prime importance. An advisory UNESCO delegation visited the lake in 1990 and pronounced favourably, although advising postponement of Baikal's nomination until some of its problems are solved. Hopefully this will bring additional pressure to bear and then an international umbrella, although Baikal's management will remain a Soviet responsibility.

Fourthly, Professor Mikhail Grachev, the dynamic new director of the Limnological Institute, founded in 1990 a Baikal International Centre for Ecological Research (BICER) to study the lake with an initial five million roubles from the Siberian Academy of Sciences. It has already attracted many foreign scientists to collaborative research ventures[44] and although conservation has not yet figured in the joint research so far, the concentration of so much international expertise must help Baikal's problems in the long run.

Lastly is the new force of 'people power' – public opinion with the clout of demonstrations and multi-choice political votes. There is much feeling about the lake, and the local greens have now formed several groupings for its defence, among them the Baikal Fund and (in Buryatia) Baikal Eco-World. Already there have been demonstrations in the streets of Irkutsk. When work was begun on a highly controversial and costly pipeline (before a formal go-ahead) to direct the effluence of Baikal'sk into the nearby Irkut River, thus maintaining the plant on the lakeside *and* ruining a popular holiday area, a demonstration of workers and students demonstrated in the path of the bulldozers and after arrests, releases and a petition which collected 104,000 signatures, the plan was dropped.

Conclusion

These are positive portents in what might seem a bleakly negative picture of the situation east of the Urals. Pristine nature still, thankfully, does exist. In Buryatia, people can still drink straight from the beautiful 12 kilometres-wide Bol'shoye Yeravnoye lake. In the same area, the upper reaches of the Udar River (badly polluted further downstream after flowing through Ulan-Ude) are crystal clear just as are the rivers of the Central Siberian zapovednik further west.[45] This nature reserve (with internationally recognised biosphere status) is by far the largest in the USSR at 1.5 million hectares. But it is only one of the twenty-odd nature reserves east of the Urals, protecting the most important habitats and species. One huge *zapovednik*, for instance, was recently established in the Lena delta as a protected breeding place for walrus, polar bear, Bewick's swan, Ross's gull, etc.[46]

The long era of grandiose schemes is, one hopes, at an end. The world's largest irrigation project, to turn the waters of Siberian rivers southwards to irrigate Central Asia, was dropped after years of debate and in-fighting, with many scientists greatly perturbed by the huge potential environmental damage. In this case, to over-simplify, the scientists triumphed over the ministries, although, it must be said, the Central Asian authorities are still pressing for the scheme. But it is the new force of public opinion which has shelved both the Katun and Turukhansk hydro-electric power schemes in Siberia at least temporarily.[47]

In the ten years up to 1989 the state spent 70 billion roubles on environmental protection (an eighth, approximately, of the US figure), mostly on purification facilities,[48] although there is a long way to go still. Another hopeful sign is that the state committee for the environment, Goskompriroda, has, as of March 1990, been changed to a ministry and its chairman to formal ministerial rank, perhaps indicating that the government intends to upgrade environmental problems.

Glasnost has educated the public in the scale and seriousness of the situation but the media has yet to educate the people as to their own environmental responsibilities. Certainly the major culprit has been industry but the individual is, in very general terms, still environmentally illiterate and irresponsible, just as we were in the West until recently.

Certainly the Soviet environmental crisis is part of a global problem with some notorious names: not only Chernobyl and the Aral Sea in the USSR, but Minimata in Japan, Bhopal in India, Seveso in Italy, etc. We have to put our own houses in order. In the case of Britain we must cease, for instance, to send acid rain to Scandinavia and to contaminate the Irish Sea with radioactive waste. With our more advanced technologies, democratic processes and other advantages, are we not more culpable than the Soviet Union with its manifest – and manifold – problems?

The root causes of the Soviet situation include the old attitude of limitless resources there to be exploited; economic self-justification of every territory; the command economy and state planning from a distant capital with the fulfilment of the state plan as the chief goal; political ambition and infighting; a plethora of uncoordinated ministries and government departments; the massive inefficiency of the system; lack of accountability and personal responsibility; and, until perestroika, the party's total control of public opinion.

In this present period of increasing economic and social chaos, including the spectre of mass unemployment, the environment is unlikely to be seen as a priority. Assuming they could get the necessary funds, antiquated factories would probably opt for new technology rather than better purification equipment. Opting for the latter, often unavailable in the USSR, can necessitate finding scarce foreign currency, although one obvious solution to this impasse is the barter deal, with industries trading raw or processed commodities. But new, conservative-sponsored legislation may now make this illegal.

Even with the most favourable political outcome it is likely to take many years to solve the country's major environmental problems. Yet, apart from all other reasons, their solution is politically important. For they have added significantly to the nation's ferment, focusing hostility to party, government and the Russian people, who are all too commonly perceived as exploiters and despoilers. Action must be seen to be taken and environmental policies be effectively applied and legally enforceable. In the long run it is the people's will (now at last operable) which can stop the ruination of their habitat and health and restore the planet's largest country to environmental harmony.

Perhaps the most basic factor is education, even if a long-term programme. It is growing, as is the environmental movement and general public concern. Excellent posters, for instance, already spread the gospel. One of them depicts the magnificent rock columns which rise above the Lena River for many miles, and bears the inscription, 'All of us now living answer for nature before our descendants and before history'.

Notes

1. *Argumenty i fakty*, no. 45 (11–17 November 1989), p. 3, via *Current Digest of the Soviet Press*, XLI, no. 50 (1989).
2. *Izvestiya*, 20 April 1975, condensed text, quoted via CDSP, 14 May 1975.
3. *Geografiya i prirodnye resursy*, no. 4 (1984).
4. *Soviet Weekly*, October 1979.
5. *Ekonomika i organizatsiya promyshlennogo proizvodstva*, no. 11 (November 1981), via CDSP, XXXIII, no. 51 (1981), pp. 12–13.
6. *Report on the USSR*, Radio Liberty, 2 September 1988, p. 14.
7. *Asian Wall Street Journal*, 23 December 1988, and 3 January, 1989, via *Oryx*, 23, no. 3 (July 1989) and *SUPAR Report*, no. 7 (July 1989), p. 87.
8. CDSP, XLI, no. 10 (1989).
9. *Moskovskiye novosti*, no. 9 (4 March 1990), abstract via CDSP XLII, no. 11 (18 April 1990).
10. *Doklad sostoyaniye prirodnoi sredy v SSSR 1988 godu*, Gosudarstvennyi komitet po okhrane prirody, Moscow 1989.
11. Rupert Cornwall, *The Independent*, 17 November 1989.
12. *Ibid.* and *SUPAR Report*, no. 7 (July 1989).
13. Alan Saunders, 'Poisoning the Arctic Skies', *Arctic Circle*, September/October 1990, pp. 22–31.
14. Andrew R. Bond, 'Air pollution in Noril'sk: a Soviet worst case?', *Soviet Geography*, XXV, no. 9 (1984), pp. 665–80.
15. Alexander Isayev, head of the State Forestry Committee, in *Soviet Weekly*, 18 November 1989.
16. Saunders, 'Poisoning the Arctic skies', p. 27, and Howard Weaver, 'Ghosts of the Gulag', *Arctic Circle*, 1, no. 2 (1990).
17. Ibid.
18. G. E. Shaw, 'Evidence for a central Eurasian source area of Arctic haze', *Nature*, 28 October 1982, pp. 815–18.
19. Personal observations and conversations, Ulan Ude, Irkutsk, etc., 1989/90.
20. *Izvestiya*, 10 November 1989, via CDSP, XLI, no. 45 (1989).
21. *Selected Environmental Issues in the Soviet Union*, prepared for Survival Anglia by Graham Drucker for IUCN East Europe Programme and World Conservation Monitoring Centre, January 1989. Unpublished report.
22. *Selected Environmental Issues*, ibid.
23. *Ekonomika i organizatsiya promyshlennogo proizvodstva*, no. 3 (March 1982) via CDSP, XXXIV, no. 23 (7 July 1982).
24. *Pravda*, 11 September 1985 via CDSP, XXXVII, no. 37 (1985).
25. *Ekonomika i organizatsiya promyshlennogo proizvodstva*, no. 3 (March 1982) via CDSP, XXXIV, no. 23 (7 July 1982).
26. V. C. Kirpichnikov, 'Sud'ba Kamchatki v nashikh rukakh!', *Priroda*, November 1990, pp. 39–46.
27. Judith Perera and Roger Milne, 'Soviet Union comes clean on nuclear blast', *New Scientist*, 5 August 1989.
28. Zhores Medvedev, personal communication, 13 March 1991.

29. Zhores Medvedev, 'Bringing the skeleton out of the closet', *Nuclear Engineering International*, November 1990, pp. 26–32.
30. *Report on the USSR*, Radio Liberty, 31 August 1990.
31. *Sotsialisticheskaya industriya*, 13 May 1989, p. 2, FBIS-Sov, 89/96, 19 May 1989, p. 88, via *SUPAR Report*, no. 7 (July 1989), p. 87.
32. *Izvestiya*, 27 January 1989, p. 2, *Pravda*, 1 February 1989, via CDSP, XLIV, no. 4 (22 February 1989), pp. 20–1.
33. AFP, 7 and 12 March 1989, *Sovetskaya Rossiya* 7 March 1989, p. 2, via CDSP, XLI, no. IX (29 March 1989), p. 21.
34. *Komsomol'skaya pravda*, 15 June 1989, p. 1, FBIS Sov, 89/118, 21 June 1989, p. 72 via *SUPAR Report*, no. 7 (July 1989).
35. Perera and Milne, 'Soviet Union comes clean'.
36. *Soviet Weekly*, 9 December 1989.
37. *Moscow News*, no. 34 (27 August/3 September 1989), p. 5, via *SUPAR Report*, January 1990, p. 90.
38. For a summary, see John Massey Stewart, '"The great lake is in great peril"', *New Scientist*, 30 June 1990, Vol. 126, no. 1723, pp. 58–60.
39. See V. V. Vorob'yov, 'Problems of Lake Baikal in the current period', and (with A. V. Martynov), 'Protected areas of the Lake Baykal basin', *Soviet Geography* (January and May 1989), XXX, nos. 1 and 5, pp. 33–48 and 359–70, translated from *Geografiya i prirodnye resursy*, nos. 3 and 2 (1988).
40. *Izvestiya*, 26 April 1989, and CDSP, XLI, no. 18 (31 May 1989), 28/30 via *SUPAR Report*, no. 7, July 1989.
41. G. I. Galazy, *Baikal v voprosakh i otvetakh*, (Moscow: Mysl', 3rd edition, 1988).
42. Ibid., particularly.
43. Vorob'yov, 'Problems of Lake Baykal'.
44. John Massey Stewart, 'Baikal's hidden depths', *New Scientist*, 23 June 1990, vol. 126 no. 1722, pp. 42–6.
45. Personal observation, 1989/90.
46. *Pravda*, 6 April 1986, p. 6, via CDSP, XXXVIII, no. IX (7 May 1986).
47. *Pravda*, 3 January 1989, p. 2, via CDSP, XLI, no. 1 (1989), p. 29.
48. *Argumenty i fakty*, no. 13 (1–7 April 1989), via CDSP, XLI, no. 18 (1989).

Index

Abalkin, Leonid, 209
Academy of Sciences, USSR, 129, 131, 137
acid rain
  in European Russia, 58
  at Lake Baikal, 232
  in Siberia, 225, 227
  and transboundary air pollution, 120
Adamov, E., 181
Aganbegyan, Abel, 224
agencies, environmental, xv, 70–1, 234
agricultural pollution
  at Lake Baikal, 231
  US–USSR environmental agreements projects, 129, 143
air pollution
  in Armenia, 16, 26–7
  in Azerbaidzhan, 16, 26, 27
  in the Baltic republics, 15–16
  in the BAM railway zone, 45, 50, 224
  in the Caucasus, 26–7
  in cities, 202, 20–5, 225–7
  and global climatic change, 119
  at Lake Baikal, 232
  official reports on, 204–5
  by republic, 16
  in Siberia, 32, 225–7
  transboundary, 120–2
  in the Ukraine, 28
  US-USSR environmental agreements projects, 129, 131, 141–2
All-Russian Society for the Conservation of Nature, 64
Amu Dar'ya River, 90, 91, 92–3, 94, 95, 97, 98, 100, 101, 102-3, 104, 105, 107
Amur River, 228
anaemia in children, 26, 27
Apatity (town), 215
Aral Sea, 8, 25–6, 33, 60–1, 88, 98–108, 122, 199, 200, 207
  ameliorative measures, 103–8
  causes of reduced inflow, 100

environmental/ecological problems, 101–3
  and the Siberian water diversion project, 105–7, 234
  water resources from, 89–90
  water use, 91, 92–3
Arctic area, 213–22
Arctic ecosystems
  US-USSR environmental agreements projects, 149
Armenia, 206
  air pollution, 16, 26–7
  environmental opposition, 6, 7
  Lake Sevan, 6, 7
  US-USSR environmental agreements projects in, 131
Arnott, Don, 183
atmospheric warming
  and the Baltic republics, 15–16
Austria
  nuclear power plants, 188
automobiles *see* cars
Azerbaidzhan
  air pollution, 16, 26, 27
  US-USSR environmental agreements projects in, 131

Baikal, Lake, 5, 30–1, 45, 46–7, 200, 227, 230–3
Baikal International Centre for Ecological Research (BICER), 233
Baikal'sk Pulp and Paper mill, 47, 67, 231–2
Baltic republics, 206
  environmental problems, 29–30
  nationalism and environmentalism, 11–23
  see also Estonia; Latvia; Lithuania
Baltic Sea, 7, 14–15
BAM (Baikal-Amur Mainline) railway, 40–56, 223–4, 232
  and human ecology, 45–51

Index

and physical ecology, 41–5
Bater, James, 54
Belgium
  nuclear energy, 187–8
Byelorussia
  air pollution, 16
  health and environmental effects of Chernobyl, 160, 161, 164
  US-USSR environmental agreements projects in, 131
Belyaev, S. T., 189
biological consequences of environmental pollution, US-USSR environmental agreements projects, 129, 146–7
Birlik (nationalist movement), 8, 26
Black Sea, 117
Bondar, Alexander, 52
Borovoi, A. A., 189, 190, 192, 193, 194
Bratsk, 228
Brezhnev period
  and the BAM railway, 41
  and environmental politics, 65, 67–9
  Nixon-Brezhnev agreements (1972), 126, 127
Britain
  and the Chernobyl accident, 177, 180, 181–3, 185–6
  environmental problems, 234–5
  and transboundary air pollution, 120
Bulgaria, nuclear energy in, 187
bureaucratic restructuring, and environmental politics, 83
Bush-Gorbachev Summit (1990), 151, 163

Canada, 220
cancer rates
  in the Aral Sea area, 103
  in Siberia, 225, 230
carbon dioxide emissions, international environmental cooperation on, 119–20
cars, pollution from, in the BAM railway zone, 51
Caspian Sea, 27
Caucasus, US-USSR environmental agreements projects in, 131
Central Asia
  birthrate, 88–9
  disappearing Aral Sea, 25–6
  ecological disaster zones, 60–2
  emigration from, 108
  environmental politics, 7–8, 71, 73–4
  population, 88-9, 90, 108
  water management, 33, 88–114, 234

Central Siberian Zapovednik, 234
CFCs, international cooperation on banning of, 118
Chara River, 48
Cherkassy, chemical pollution in, 28
Chernobyl accident, 122, 199
  and the Baltic republics, 14
  economic costs of, 195
  and environmental politics, 65, 66
  and glasnost, 4
  global impact of, 174–96
  and the international nuclear industry, 184–8
  and the sarcophagus, 188–94
  Soviet public opinion on, 164–5
  and the Ukraine, 8-9, 28, 29
  and US-Soviet nuclear safety cooperation, 150–1, 152, 153–7, 159, 160–4
  Western reconstruction of, 177–83
Chernobyl Notes (Medvedev), 180, 192–3
children, and the Chernobyl accident, 161
children's illnesses
  in Azerbaidzhan, 27
  in the BAM railway zone, 53
  beyond the Urals, 227
  in Central Asia, 26
  in Estonia, 18
China, People's Republic of, and industrial carbon dioxide discharges, 119
Chita, 52
Chukchi people, 230
Chukotka Peninsula, 230
cities, and air pollution, 202, 204–5, 225–7
citizens' groups, and environmental politics, 72
class politics, and environmental politics, 77, 78
climatic changes
  in the Aral Sea area, 103
  global, 117, 119–20
  US-USSR environmental agreements projects, 129, 147
coastline pollution, in the Baltic republics, 14–15, 29
COMECON (Council for Mutual Economic Aid), 208
concepts, substitution or misuse of, in the Soviet Union, 202
conflict, and environmental politics, 64, 65, 84
Congress of Peoples' Deputies, and environmental politics, 70, 73, 81, 82

conservation societies, and environmental politics, 74–5
Cooperative Science Programme (US-USSR), 126
corporatist structures, formalised, and environmental politics, 80, 83–4
cotton cultivation, in Central Asia, 33, 89, 91, 97–8
cotton monocultures, in Central Asia, 8, 199
CPSU (Communist Party of the Soviet Union), 210
  and environmental politics, 66
  and the northern peoples, 220–1
  and the Siberian water diversion project, 106
crop production, in Central Asia, 89, 91, 94, 97–8
cultural values, and environmental politics, 76–7
Czechoslovakia
  and nuclear energy, 187
  and transboundary air pollution, 120

Davydov project, 105
death rates in Siberia, 230 *see also* infant mortality
decision-making processes, in the Soviet Union, 197, 207–8
desertification, 60, 61–2, 102
disasters *see* environmental disasters
diseases *see* illnesses
Dneiper River, 200
drinking water, in the Aral Sea area, 103

earthquake prediction, US-USSR environmental agreements projects, 129, 131, 148
earthquakes, in the BAM railway zone, 45
Easton, John, 163
eco-sovietology, 197–212
  research on, 210–11
ecological disasters *see* environmental disasters
economic independence, in the Baltic republics, 21–2
ecosystems, degradation of, 57–63
ecumenes, 40, 51
EEC (European Economic Community), 118, 120
energy industry, and transboundary air pollution, 121
energy sources
  alternative, 120

shift in, and global warming, 120
  in Siberia, 213, 218, 219
  see also natural gas; oil
Environmental Business Association, 75
environmental disasters, 57, 58–60
  areas of, 58–62, 200
environmental laws
  lack of enforcement, 3
  in the West, 3
environmental management, by the state, 123–4
Estonia
  air pollution, 16
  coastline pollution, 14–15
  environmental opposition, 7
  Green Movement (EGM), 19
  mining activities, 16-18, 20, 22
  nature reserves, 11, 12, 13
  US-USSR environmental agreements projects in, 131, 137
exchanges, and the Environmental Agreement projects, 135–6

families, in the BAM railway zone, 54
Far East (Soviet), 31, 54
  and the BAM railway, 41
  housing, 53
  see also Siberia
Far North (Soviet), 213–22
  ecosystem degradation, 62
  population, 54
farms, and water pricing, 98
FESCO (Far Eastern Shipping Company), 229–30
Fevral'sk (city), 44
Finland
  nuclear energy, 187
  and transboundary air pollution, 120, 121
Finland, Gulf of, and the Leningrad flood control dike, 19, 32–3
fish, destruction of
  in the Aral Sea, 102
  in the Soviet Arctic, 215, 225, 228, 232–3
food production, in Central Asia, 91, 94, 97, 109
food shortages, in the BAM railway zone, 52–3
forests
  in the Baltic republics, 12
  in the BAM railway zone, 45–6, 49
  taiga, 223, 224
France, and the Chernobyl accident, 177, 186, 187

Index

GAEN (Soviet Nuclear Regulatory Commission), 157, 159
Gagarin, Yuri, 52
Gagarinsky, Yu., 189
Galazy, Grigory, 231
gas see natural gas
genetic consequences of environmental pollution, US-USSR environmental agreements projects, 129, 146–7
geographical conditions, in Siberia, 31
Georgia
  air pollution, 16
  US-USSR environmental agreements projects in, 129, 131, 132
Germany
  nuclear energy, 188
  and transboundary air pollution, 120
Gilyui River, 50
glasnost
  and eco-sovietology, 197–212
  ecological, 65, 66
  effects of, 1
  and the environment, 4–5
  and environmental politics, 79–80
  and nationalist movements, 2
Glinka, M. I., 40, 41
global environmental changes, 117–22
Gorbachev, Mikhail, 209
  and the BAM railway, 41
  Bush-Gorbachev Summit (1990), 151, 163
  reforms, 211
Gorbachev period, and environmental politics, 65–7
Gorbunov, V. A., 49
Gorsuch, Ann, 127
Goskompriroda (State Committee for Environmental Protection), 66, 70–1, 73, 75, 116, 127, 137
  report on the environment, 202–5
Grachev, Mikhail, 233
green movement, 10
  in Estonia (EGM), 19
  in Moldavia, 27–8
  in the Soviet Arctic, 226–7
Gusinoye, Lake, 228

health effects of pollution
  of air pollution, 204, 205
  of the Chernobyl accident, 152, 155, 160-4
  public attitudes to, 34–5
health services, in the BAM railway zone, 53
housing, in the BAM railway zone, 53

human values, and environmental politics, 76–7
hydro-electric projects, 228
Hydromet (State Committee for Hydrometeorology and Environmental Control), 127, 137

IAEA (International Atomic Energy Agency), 151, 152, 153–4, 155–6, 160, 164–7, 176, 177–83, 189–90, 191–2
Il'ina, L. N., 45–6, 47
illnesses
  and air pollution, 205
  and the Aral Sea desiccation, 102, 103
  in Siberia, 230
  in the Soviet Arctic, 215
  see also children's illnesses
ILO (International Labour Organisation), 213–14
industrial development
  nature-destroying, 203–4
  in the Soviet Arctic, 215
industrial pollution, economic damage from, 115
industrial workers, and environmental politics, 74
infant mortality
  in the Aral Sea area, 103
  in Central Asia, 26
  of northern peoples, 216
  in Siberia, 223
information, lack of, for eco-sovietologists, 201–2
intelligentsia, and environmental politics, 64, 74
interest group politics, 80–1
international cooperation, on environmental protection, 115–24
International Environmental Security (IES), 115–16
irrigation
  and the Aral Sea, 100–1
  in Central Asia, 33, 88, 91, 94, 94–8, 234
Isayev, A. S., 228
Italy, and nuclear power plants, 188
Izrael, Yu., 205

Japan
  and the Chernobyl accident, 177, 184, 187
  nuclear energy, 187
  and ozone-depleting substances, 118

Index

JCCCNRS (Joint Coordinating Committee on Civilian Nuclear Reactor Safety), 156, 157, 158, 163, 166
Joint Agreement on the Environment, 125–49

Kaipbegenov, Tulepbergen, 107–8
Kalmyk steppes, desertification in, 62
Kamchatka Peninsula, 228–9
Kara-Kum Canal, 100–1
Karakalpak ASSR, 103
Karimov, I. A., 106
Kazakhstan, 9, 199
  air pollution, 16
  area and population, 90
  ecosystem degradation, 61–2
  US-USSR environmental agreements projects, 131, 132
  water management, 106
Kemerovo (city), 204
Kennan Report (1985), 128
Khanka, Lake, 229
Khudonazarov, Davlat, 8
Kirgizia
  air pollution, 16
  area and population, 90
  US-USSR environmental agreements projects in, 131
  water management, 94, 106
Kola Peninsula, 215
  transboundary air pollution, 120, 121–2
Komarov, Boris, 198–9
Konovalov, Vitaly, 163
Koryakin, Yu. I., 195
Kotlyakov, V., 220
Krasnoyarsk, 230
Kurchatov, I., 185
Kyshtym, 229

Ladoga, Lake, 32
land value, assessment of, in the Soviet Arctic, 220
Lapidus, Gail, 79
Lapshin, Alexander, 156
Latvia
  air pollution, 16
  coastline pollution, 14, 15, 29
  environmental groups, 7, 19
  environmental problems, 29–30
  Gauya National Park, 12–13
  Museum of Nature in Riga, 20
  nature reserves, 11, 12, 13
  US-USSR environmental agreements projects in, 131

Legasov, Valery, 4, 178
legislation
  environmental, 123
  and environmental politics, 65, 83
  on the Soviet Arctic, 218–19
Lemeshev, M., 200
Lena River, 234, 235
Leningrad, as disaster area, 58
life expectancy
  of northern peoples, 216–17
  in Siberia, 225, 226
  in the Soviet Union, 204
Lithuania
  air pollution, 15, 16
  coastline pollution, 15
  environmental groups, 7, 19
  Ignalina nuclear power complex, 7, 13, 14, 19, 209
  National Park, 13
  nature reserves, 12, 13
  Sajudis movement, 209
  US-USSR environmental agreements projects in, 131
living standards, of northern peoples, 217
local organisations, and environmental politics, 71–2, 77–8
Lukonin, Nikolai, 157

Malyshev, Vadim, 156–8
maps, and the Goskompriroda report, 202, 203
marine pollution
  in Siberia, 227
  US-USSR environmental agreements projects, 129, 146
market reforms, and environmental politics, 74, 75, 82
Marshall, Lord, 177, 182, 186
material values, and environmental politics, 76, 77
Matusevich, Vladimir, 209–10
media, the, and environmental politics, 69, 72–3
Medvedev, Grigory, 180, 192–3
Medvedev, Zhores, 229
Mel'nikov, Vladimir, 224
migration
  to the BAM railway zone, 53
  to Siberia, 40–1
military zones, 205
military-industrial complex, in the Soviet Union, 207
mining activities, in Estonia, 16–18
Mogilev (city), 204

Moldavia
  air pollution, 16
  Green movement, 27–8
  US-USSR environmental agreements projects in, 131
Monin, A., 61
Montreal Protocol (1987), 118
Morgun, Fyodor, 18, 138
Muya Tunnel, 44

Nagorno-Karabakh, 7
nationalism, 1, 2
  environmentalism, and political participation, 35–6
natural gas
  in Siberia, 213, 214–15, 218, 219, 227
  in the Soviet Arctic, 214–15, 218, 219
natural resources, in the Soviet Union, 57
nature
  conquest of, in Siberia, 223–5
  preservation of: US-USSR environmental agreements projects, 129, 144–6
nature reserves
  in the Baltic republics, 12–14
  in Siberia, 234
  in the Soviet Arctic, 219
Nentsy people, 230
newspapers, and environmental politics, 72
NIMBY ('Not in My Back Yard') attitude, 35, 81
nitrogen oxide emissions, and transboundary air pollution, 120, 121
Nixon-Brezhnev agreements (1972), 126, 127
Norilsk, 32, 205, 215, 226–7
northern eco-systems
  fragility of, 197
  see also Far North (Soviet)
northern peoples, 213–14
  and economic development, 220–1
  life-style, 216-17
Novokuznetsk (city), 225–6, 227
Novosibirsk (city), 228
  air pollution, 32, 204, 225
nuclear pollution, beyond the Urals, 229–30
nuclear power plants, 4, 120, 199
  and environmental politics, 65
  Ignalina (Lithuania), 7, 13, 14, 19, 209
  international cooperation on safety, 151–73
  in other countries, 184-8
  safety of, after Chernobyl, 174-96

situation of, 4
  see also Chernobyl accident
nuclear weapons production, 207
nuclear weapons testing
  resistance to, 9
  in the Udokan copper basin, 48

Ob' river, 228, 229
oil
  and nuclear energy, 188
  in the Soviet Arctic, 213, 218, 219
oil spills
  in Siberia, 227
  in the Soviet Arctic, 215
organisations, environmental
  ad hoc independent public, 80–1
  and glasnost, 4–5, 6
  local, and environmental politics, 71–2, 77–8
  unofficial, 73–5
Ovsyannikov, Nikolai, 3
ozone layer, depletion of, 117, 118

Pamyat', 9, 74
perestroika
  and the Baltic republics, 21
  and eco-sovietology, 197–212
  and international environmental cooperation, 115–24
permafrost, 223, 224
  and the BAM railway, 43–4
policy choices, and environmental politics, 82–3
political participation, and environmental issues, 24–5, 33–6
politics, environmental, 64–87
pollution, changing attitudes to, 1–2
pollution control, Soviet interpretation of, 202
popular front movements, 206–7
Preobrazhensky, 57
pricing, irrigation-water, 98
protests, environmental
  and the Chernobyl accident, 151
  growth of, 5–6
  of northern peoples, 216
  in the pre-Gorbachev era, 3
Pryde, Philip, 138
public opinion, 124
  and environmental problems, 235
  and the health effects of pollution, 34–5
  and Lake Baikal, 233
  and the Soviet nuclear power programme, 164–5

PWRs (pressurised water reactors), 155, 175–6, 184

quality of life, in the BAM railway zone, 52–4

radiation
  and the Chernobyl accident, 161, 162, 164, 194
  and ecosystem degradation, 58–9
  in Siberia, 214, 230
RAMSAC Convention, 229
Reilly, William, 127
reindeer husbandry, in the Soviet Arctic, 214, 216, 217
research, environmental
  eco-sovietological, 210–11
  in the Soviet Arctic, 218
Reteyum, A., 61
river diversion schemes, Siberia, 33, 105–7, 109–10, 234
rivers, pollution of, in Siberia, 227–9
Robinson, N., 128, 137
Russian republic (RSFSR)
  air pollution, 16
  and the Aral Sea problem, 107, 109
  and environmental politics, 30–3, 77
  health and environmental effects of Chernobyl, 160
  housing, 53
  Ministry of Land Reclamation and Water Management, 3
  nationalism and environmentalism, 9–10
  US-USSR environmental agreements projects in, 131, 132

Sakhalin Island, 228
SANIIRI (Central Asian Institute for Irrigation Research), 95
Scientific-Technical Council, 70
Selenga River, 233
Semeldzha River, 44
Sevan, Lake, 6, 7
Severobaikal'sk (city), 46–7, 50, 232
Seveso, 234
sewage treatment, in the BAM railway zone, 49–50
Siberia, 223–37
  air pollution, 225–7
  conquest of nature in, 223–5
  environmental problems, 31–3
  migration to, 40–1
  nuclear pollution, 229–30
  water diversion project, 33, 105–7, 109–10, 234

water problems, 227–9
  see also Far East (Soviet); Far North (Soviet)
Social and Ecological Union, 81
Social-Ecological Union, 74
Societal Council, 70
Society for the Protection of Nature, 3, 5
soil degradation/erosion, 58, 199
Sokolovsky, Valentin, 127, 203
South Korea, and nuclear energy, 187
Spain, and nuclear energy, 188
Stalin, Joseph, 7, 223
subarctic ecosystems, US-USSR environmental agreements projects, 149
sulphur deposition, in the Soviet Arctic, 215
sulphur dioxide emissions, and transboundary air pollution, 120, 121
Supreme Soviet, and environmental politics, 73, 82
Sweden, and nuclear energy, 188
Switzerland, and nuclear energy, 188
symbolic significance of politics, 78–80, 83
Syr Dar'ya River, 90, 91, 92–3, 94, 95, 97, 98, 100, 102–3, 104, 105, 107

Tadzhikistan
  air pollution, 16
  area and population, 90
  cotton monoculture, 8
  US-USSR environmental agreements projects in, 129, 131, 132, 137
  water management, 94, 106
taiga, 223, 224
Taimyr Green Front, 227
Taimyr Peninsula, 225
Taiwan, nuclear energy, 187
Tarakanov, Nikolai, 189
Tengiz area, ecosystem degradation, 61–2
TerKSOP (Comprehensive Territorial Environmental Plan), 233
Three Mile Island, see under United States
timber industry
  and the pollution of Lake Baikal, 230–1
  and the pollution of Siberian rivers, 228
Tomsk nuclear power station, 229
Treadgold, Donald W., 40
tundra, 223, 224, 225
  ecosystem degradation, 62
Turkmenia
  air pollution, 16
  area and population, 90
  infant mortality, 103

Index

US-USSR environmental agreements
    projects in, 131
    water management, 94, 106
Tynda (city), 50–1, 53

Udokan copper basin, 47–9
Ukraine
    air pollution, 16
    and the Chernobyl disaster, 8–9, 28, 29
    environmental problems, 28–9
    health and environmental effects of Chernobyl, 160, 164
    US-USSR environmental agreements projects in, 131, 132
Ulan-Ude, 227, 232, 234
unemployment, in the BAM railway zone, 52
UNEP (United Nations Environment Programme)
    and the Aral Sea problem, 106–7, 108
United Nations
    Conference on Environmental Development (1992), 116
    and international environmental cooperation, 116
United States
    and air pollution, 122
    and the Chernobyl accident, 177–8, 179-80, 183, 187, 188, 195
    Environmental Protection Agency (EPA), 127, 137
    environmental spending, 123
    industrial carbon dioxide discharges, 119
    Joint Agreement on the Environment (1972), 125–49
    northern territories, 220
    nuclear power plants, 184, 185, 187, 188
    Nuclear Regulatory Commission (NRC), 150, 152–3, 154, 158–9
    and nuclear safety cooperation, 150–73
    and ozone-depleting substances, 118
    Reagan administration, 127, 152–3
    Three Mile Island nuclear accident, 154, 157, 174, 175, 176, 184, 188, 194–5
    US-Soviet Joint Committee (JCM), 153, 154, 163
urban environment
    US-USSR environmental agreements projects, 129, 132, 143–4
Ust' Ilimsk, 231
Uzbekistan
    air pollution, 16

area and population, 90
environmental issues, 26
US-USSR environmental agreements
    projects in, 131
    water management, 94, 98, 106

vegetation, in the Aral Sea area, 102
Velikhov, Evgeny, 163
Vinogradov, Boris, 200
Volga delta, ecosystem degradation, 59, 62
Vorob'yov, V. V., 231, 233
Vorontsov, Nikolai, 70, 200, 201, 204

wages
    of northern peoples, 217
water management
    in Central Asia, 88–114
    ineffectiveness of, 122–3
water pollution
    in the Ukraine, 28–9
    US-USSR environmental agreements projects, 129, 131, 142–3
water pricing, in Central Asia, 98
water problems, in Siberia, 227–9
water supplies, in the BAM railway zone, 49–50
Watkins, James, 163
Waxmonsky, G., 128, 137
West, the, and environmental politics, 81
Western Europe, industrial carbon dioxide discharges, 119
wildlife
    in the Aral Sea area, 103
    in the Baltic republics, 14
    in Siberia, 234
    US-USSR exchange of, 128
Wolfson, Zeev, 198

Yablokov, Aleksei, 32, 73, 200
Yamal Peninsula, 62, 218, 219, 224–5
Yamal-Nenets Autonomous District, 213, 214
Yavlinsky, Grigory, 209
Yenisei River, 228

Zabelin, S., 61
Zade, Polad-Polad, 96
Zalygin, Sergei, 32
zapovedniki (state nature reserves), in the Baltic republics, 12–14
Zech, Lando, 156, 157
Zeya River, 50
Ziegler, Charles, 128
Zimnina, T. I., 46
ZumBrunnen, Craig, 128

SELECTED PAPERS FROM THE FOURTH WORLD CONGRESS FOR SOVIET AND EAST EUROPEAN STUDIES, HARROGATE, JULY 1990

Edited for the International Committee for Soviet and East European Studies by Stephen White, University of Glasgow

Titles published by Cambridge

Market socialism or the restoration of capitalism?
edited by ANDERS ÅSLUND

Women and society in Russia and the Soviet Union
edited by LINDA EDMONDSON

Soviet foreign policy in transition
edited by ROGER KANET, DEBORAH NUTTER MINER and TAMARA J. RESLER

The Soviet Union and Eastern Europe in the global economy
edited by MARIE LAVIGNE

The Soviet environment: problems, policies and politics
edited by JOHN MASSEY STEWART

New directions in Soviet history
edited by STEPHEN WHITE

www.ingramcontent.com/pod-product-compliance
Ingram Content Group UK Ltd.
Pitfield, Milton Keynes, MK11 3LW, UK
UKHW040704180125
453697UK00010B/403